COMPLETE GUIDE TO MODERN VCR TROUBLESHOOTING AND REPAIR

OTHER BOOKS BY JOHN D. LENK

(160813) COMPLETE GUIDE TO LASER/VIDEODISC PLAYER TROUBLESHOOTING AND REPAIR, $27.95 (1985)
(160820) COMPLETE GUIDE TO VIDEOCASSETTE RECORDER OPERATION AND SERVICING, $27.95 (1983)
(372391) HANDBOOK OF ADVANCED TROUBLESHOOTING, $24.95 (1983)
(377317) HANDBOOK OF DATA COMMUNICAITONS, $24.95 (1984)
(380519) HANDBOOK OF MICROCOMPUTER BASED INSTRUMENTATION AND CONTROLS, $24.95 (1984)
(381666) HANDBOOK OF SIMPLIFIED COMMERCIAL AND INDUSTRIAL WIRING DESIGN, $24.95 (1984)
(392473) A HOBBYIST'S GUIDE TO COMPUTER EXPERIMENTATION, $24.95 (1985)

Order from: Steven T. Landis
College M. O. Sales
Prentice-Hall, Inc.
200 Old Tappan Road
Old Tappan, New Jersey 07675

COMPLETE GUIDE TO MODERN VCR TROUBLESHOOTING AND REPAIR

JOHN D. LENK

Consulting Technical Writer

Prentice-Hall, Inc., Englewood Cliffs, New Jersey 07632

Library of Congress Cataloging in Publication Data

Lenk, John D. (date)
　Complete guide to modern VCR troubleshooting and repair.

　Includes index.
　1. Video tape recorders and recording—Maintenance and repair.　I. Title.
TK6655.V5L458　1985　　　621.388′332′0288　　　84-26266
ISBN　0-13-160359-0

Editorial/production supervision: **Lisa Schulz**
Interior design: **Judy Matz-Coniglio**
Cover design: **Photo Plus Art**

©1985 by **Prentice-Hall, Inc.**, Englewood Cliffs, New Jersey 07632

All rights reserved. No part of this book may be reproduced, in any form or by any means, without permission in writing from the publisher.

Printed in the United States of America

10　9　8　7　6　5　4　3　2

ISBN　0-13-160359-0　01

PRENTICE-HALL INTERNATIONAL, INC., *London*
PRENTICE-HALL OF AUSTRALIA PTY. LIMITED, *Sydney*
EDITORA PRENTICE-HALL DO BRASIL, LTDA., *Rio de Janeiro*
PRENTICE-HALL CANADA INC., *Toronto*
PRENTICE-HALL HISPANOAMERICANA, S.A., *Mexico*
PRENTICE-HALL OF INDIA PRIVATE LIMITED, *New Delhi*
PRENTICE-HALL OF JAPAN, INC., *Tokyo*
PRENTICE-HALL OF SOUTHEAST ASIA PTE. LTD., *Singapore*
WHITEHALL BOOKS LIMITED, *Wellington, New Zealand*

IRENE, Empress of Lake Itchygoomie

LAMBIE, Wookie of Lake Itchygoomie

KAREN and **TOM,** Queen and King of Orange

BRANDON, JUSTIN, MICHAEL, and **CATHIE**,
Princes and Princess of Orange

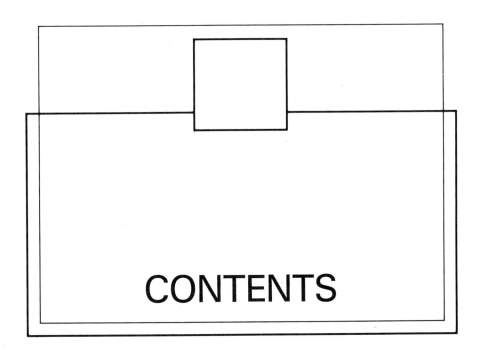

CONTENTS

Preface *xi*

CHAPTER 1   Introduction to Videocassette Recorders   *1*
   1-1   VCR basics   *1*
   1-2   Relationship of recording/playback heads   *7*
   1-3   Relationship of rotating heads to video fields and frames   *9*
   1-4   The basic VCR servo system   *10*
   1-5   Relationship of luminance and chroma signals   *14*
   1-6   Beta system basics   *15*
   1-7   VHS system basics   *30*
   1-8   The basic VCR system control functions   *45*
   1-9   Inserting and removing a videocassette   *46*
   1-10   Record lockout or malerase functions   *48*
   1-11   Features of a modern VCR   *49*
   1-12   Operating controls, indicators, and connectors of a modern VCR   *53*

CHAPTER 2   The Basics of VCR Troubleshooting and Repair   *58*
   2-1   Safety precautions in VCR service   *58*
   2-2   Test equipment for VCR service   *64*
   2-3   Tools and fixtures for VCR service   *69*
   2-4   Cleaning, lubrication, and general maintenance   *71*
   2-5   The basic troubleshooting approach   *74*
   2-6   Basic VCR troubleshooting procedures   *77*
   2-7   VCR troubleshooting/repair notes   *79*

CHAPTER 3 Power Supply Circuits 86
   3-1 Four-head VCR power supply 86
   3-2 Five-head VCR power supply 90

CHAPTER 4 Remote-Control Circuits 100
   4-1 Typical remote-control transmission formats 100
   4-2 Typical remote-control transmitter 102
   4-3 Typical remote-control preamplifier and noise filter 104
   4-4 Typical remote-control decoder microprocessor 107
   4-5 TV/VCR-select circuit 109
   4-6 Slow-tracking control generation 110
   4-7 Tuner/timer/charger communications to VCR 112

CHAPTER 5 The Mechanical Sections 114
   5-1 Typical top-load mechanical operation 114
   5-2 Typical front-load mechanical operation 123
   5-3 Mechanical adjustments 144
   5-4 Tape loading-motor and reel-motor drive circuits 144
   5-5 Cassette loading circuits 149

CHAPTER 6 Video Head Configurations and Switching Circuits 154
   6-1 Video head switching and selection concepts 154
   6-2 Four-head with 90° spacing 158
   6-3 4X video head system 161
   6-4 Five-head system 165

CHAPTER 7 Drum/Cylinder Servo 169
   7-1 Basic drum/cylinder phase control 169
   7-2 Basic drum/cylinder speed control 174
   7-3 Basic drum/cylinder motor drive 177
   7-4 Typical cylinder servo operation 180

CHAPTER 8 Capstan Servo 185
   8-1 Basic capstan servo system 185
   8-2 Capstan phase-loop operation 186
   8-3 Capstan speed-loop operation 189
   8-4 Capstan slow/still operation 191
   8-5 Using frame advance and the capstan servo in still 196
   8-6 Operating the capstan servo at double speed 199

CHAPTER 9 Special Effects 203
   9-1 Speed search 203
   9-2 Slow motion (continuously variable) 208

Contents   ix

9-3   Mechanical brake   *216*
9-4   Quasi-sync generator (substitute vertical sync)   *217*

CHAPTER 10   Tuner and Frequency Synthesis Circuits   *221*
   10-1   PLL basics   *221*
   10-2   Typical FS tuning system   *225*
   10-3   Another frequency synthesis tuner system   *236*

CHAPTER 11   Audio Circuits   *243*
   11-1   Stereo audio-head construction   *244*
   11-2   The Dolby System   *245*
   11-3   Typical audio record circuits   *246*
   11-4   Typical audio monitoring circuits   *247*
   11-5   Typical playback audio circuits   *248*
   11-6   Typical monitor select switch   *249*
   11-7   Typical audio crosstalk canceller circuits   *250*
   11-8   Audio circuit troubleshooting and repair   *251*

CHAPTER 12   Video Circuits   *253*
   12-1   Introduction to Beta circuits   *254*
   12-2   Beta luminance circuits   *255*
   12-3   Beta chroma circuits   *264*
   12-4   Luminance (black-and-white) troubleshooting/repair   *273*
   12-5   Chroma troubleshooting/repair   *274*

CHAPTER 13   Special Features and Circuits   *277*
   13-1   Typical timer and backup power supply   *277*
   13-2   An alternate timer-IC/keyboard/display configuration   *286*
   13-3   Typical electronic counter and display   *290*
   13-4   An alternate tape-counter/time-remaining configuration   *298*
   13-5   Fine editing   *303*
   13-6   Index operation   *305*

CHAPTER 14   System Control   *311*
   14-1   Microprocessor communications bus   *312*
   14-2   System-control overview   *313*
   14-3   VCR on/off operation   *318*
   14-4   System-control input operation   *320*
   14-5   Trouble sensors   *323*
   14-6   Counter/record time   *326*

Index   *329*

PREFACE

The main purpose of this book is to provide a simplified, practical system of troubleshooting and repair for the many types and models of modern videocassette recorders (VCRs). The book is the ideal companion to my best-selling *Complete Guide to Videocassette Recorder Operation and Service.* (Englewood Cliffs, N.J.: Prentice-Hall, Inc., 1983). However, there is no reference to the earlier book, nor is it necessary to have any other book to make full use of the information presented here. Of course, it is assumed that you are already familiar with the basics of television and magnetic recording. If you are not, you should not attempt troubleshooting or repair of any VCR, either modern or one of the older models.

It is virtually impossible to cover detailed troubleshooting and repair for all VCRs in any one book. Similarly, it is impractical to attempt such coverage, since rapid technological advances soon make such a book's details obsolete. Very simply, you must have adequate service literature for any specific model of VCR that you are servicing. You need schematic diagrams, part-location photos, descriptions of adjustment procedures, and so on, to do a proper troubleshooting/repair job.

Instead of trying to provide such details, this book concentrates on *troubleshooting/repair approaches.* This is done by breaking a VCR down into its various circuits or sections. All VCRs have certain circuits and/or sections in common (such as system control, video, audio, servo, etc.). A separate chapter is devoted to each of the major sections.

Using this chapter/circuit-group approach, you can quickly locate information you need to troubleshoot a malfunctioning VCR. For example, if you have audio problems, refer to Chapter 11. If the problems appear to be video-related, consult Chapter 12. In each chapter you will find: (1) an introduction that describes the purpose or function of the circuit; (2) some typical circuit descriptions or circuit theory (drawn from a cross section of VCRs); and (3) a logical troubleshooting/repair approach for the circuit (based on manufacturer's recommendations).

Chapter 1 is devoted to the basics of VCRs, both Beta and VHS. This information is included for those readers totally unfamiliar with VCRs, or those who need a refresher. This is followed by brief descriptions of the functions, features, controls, and connectors found on modern VCRs (particularly as compared to early model VCRs).

Chapter 2 describes the basics of VCR troubleshooting and repair. Such subjects as safety precautions, test equipment, tools and fixtures, cleaning, lubrication, general maintenance, a basic troubleshooting approach, basic VCR troubleshooting procedures, and VCR troubleshooting/repair notes are covered in full detail.

Chapter 3 provides an introduction to VCR power-supply circuits. This is followed by descriptions of typical VCR power-supply circuits. The chapter concludes with logical troubleshooting/repair approaches for the specific circuits.

Chapters 4 through 14 provide coverage similar to that of Chapter 3, but for such circuits or sections of a VCR as remote control, mechanical control, head arrangements, drum servo, capstan servo, special effects, tuning, audio, video, special features, and system control.

By studying the circuits and mechanical sections found in Chapters 3 through 14, you should have no difficulty in understanding the schematic and block diagrams of similar Beta and VHS VCRs. This understanding is essential for logical troubleshooting, no matter what type of electronic equipment is involved.

No attempt has been made to duplicate the full schematics for all circuits. Such schematics are found in the service literature for the particular VCR. Instead of a full schematic, the circuit descriptions are supplemented with partial schematics and block diagrams that show such important areas as signal flow paths, input/output, adjustment controls, test points, and power-source connections. These are the areas most important in troubleshooting. By reducing the full schematics to these areas, you will find the circuit easier to understand, and you will be able to relate circuit operation to the corresponding circuit of the VCR you are servicing.

Many professionals have contributed their talent and knowledge to the preparation of this book. I gratefully acknowledge that the tremendous effort to make this book such a comprehensive work would have been impossible for one person, and I wish to thank all who have contributed, both directly and indirectly.

I wish to give special thanks to the following: Bob Carlson and Martin

Plude' of B&K-Precision Dynascan Corporation; Thomas Roscoe and Eddie Motokane of Hitachi; Everett Sheppard and Jeff Harris of Mitsubishi, Deborah Fee of NAP Consumer Electronics (Magnavox, Sylvania, Philco); Thomas Lauterback of Quasar; J. W. Phipps of RCA; Donald Woolhouse of Sanyo; Richard Wheeler, J. Philip Stack, Ralph White, Jason Farrow of Sony; and John Taylor of Zenith.

I also extend my gratitude to Tim McEwen, Leon Liguori, Greg Burnell, Donna Sepkowski, Rosalie Herion, Dave Boelio, Hank Kennedy, Fay Ahuja, John Davis, Matt Fox, Barbara Cassel, Jerry Slawney, Art Rittenberg, Ellen Denning, Beverly Vill, Karen Fortgang, Lisa Schulz, Vicky Willows, Mary O'Brien, Irene Springer, Dave Amerman, and Don Schaefer of Prentice-Hall, and Marie Barlettano of PH International. Their faith has given me encouragement, and their editorial and marketing expertise has made many of my books best-sellers. I also wish to thank Joseph A. Labok of Los Angeles Valley College for his help and encouragement.

John D. Lenk

COMPLETE GUIDE TO MODERN VCR TROUBLESHOOTING AND REPAIR

1

INTRODUCTION TO VIDEOCASSETTE RECORDERS

This chapter is devoted to the basics of videocassette recorders (VCRs), both Beta and VHS. The introduction is included for those readers totally unfamiliar with VCRs, or those who need a refresher, or those few who have not yet read my best-selling *Complete Guide to Videocassette Recorder Operation and Servicing*.

1–1 VCR BASICS

A VCR is a form of VTR, or *video tape recorder*. VTRs have been used in the television industry since about 1955. They generally use larger tape formats (2-, 1-, and ¾-in.) and may or may not use cassettes. In this book I concentrate on consumer-type VCRs that use ½-in. cassettes. I also concentrate on VCRs that are compatible with the NTSC television broadcast system.

Figure 1–1 shows some modern VCRs described in this book. Although the VCRs use the same principles as an audio tape recorder, operation of a VCR is much more complex. In addition to audio information, the VCR must also record and play back video information (both black and white and color) as well as control or synchronization information.

Figure 1–2 shows the basic functional sections of the VCR, which include a tuner (UHF and VHF), an RF section, a timer section, and a mechanical section (including tape transport, stationary audio head, stationary control head, and rotating video heads). Note that the same heads used for record are also used for playback. On early-model VCRs there are two rotating heads

FIGURE 1-1a Mitsubishi HS-33OUR Four-Head Videocassette Recorder (Courtesy of Mitsubishi Electric Sales America, Inc.)

used for video record/playback. On most modern VCRs the trend is toward four heads, or possibly five heads. The four- and five-head models (which are available in more than one head configuration—90°, 180°, etc.) provide additional features, particularly with regard to high-speed recording and still-frame or slow-motion playback. We discuss these circuits in later chapters.

In consumer VCRs there is one audio head and one control head. Usually both heads are combined in one head stack, with the audio head at the top.

FIGURE 1-1b Sylvania VC3630 Stereo Videocassette Recorder (Courtesy of N.A.P. Consumer Electronics Corporation-Sylvania Audio-Video Product)

VCR Basics 3

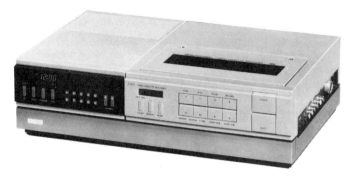

FIGURE 1-1c RCA VJP900 Convertible Videocassette Recorder (Courtesy of RCA Corporation)

FIGURE 1-1d RCA VJT400 SelectaVision Videocassette Recorder (Courtesy of RCA Corporation)

FIGURE 1-1e RCA VJT500 SelectaVision Videocassette Recorder (Courtesy of RCA Corporation)

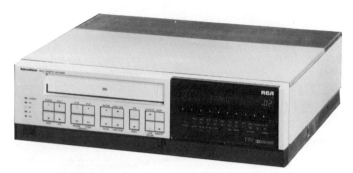

FIGURE 1-1f RCA VJT700 SelectaVision Videocassette Recorder (Courtesy of RCA Corporation)

FIGURE 1-1g Quasar VH5845XQ Stereo Videocassette Recorder (Courtesy of Quasar Company, Franklin Park, Illinois)

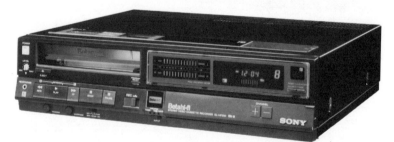

FIGURE 1-1h Sony SL-HF300B Stereo VideoCassette Recorder (Courtesy of Sony Corporation of America)

VCR Basics   5

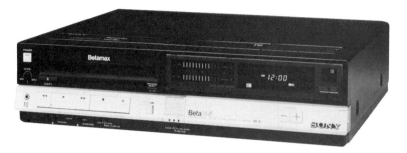

FIGURE 1-1i Sony SL-HF500 Stereo VideoCassette Recorder (Courtesy of Sony Corporation of America)

In modern VCRs the heads provide *two audio channels* for stereo or two-channel independent operation. In all VCRs (but not shown in Fig. 1-2) is a full-track erase head which erases audio/video/control information from the tape. On some modern VCRs there is also an audio-track erase head which erases either or both audio channels without disturbing the video or control information. This permits *dubbing of audio information* onto tapes recorded with video/control information.

We discuss full circuit details throughout this book. For now, let us consider the basic functions of the VCR sections shown in Fig. 1-2. The tuner section is similar to tuners found in TV sets, and functions to convert broadcast signals picked up by the antenna to frequencies and formats suitable for use by the VCR. TV channels 2 through 83 are covered by all VCRs, with additional cable TV channels available on some modern VCRs. The tuners

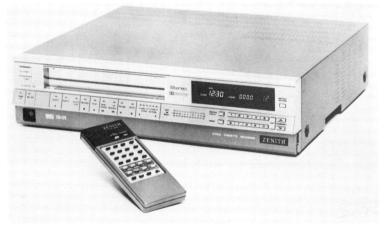

FIGURE 1-1j Zenith VR4000 Stereo Videocassette Recorder (Courtesy of Zenith Electronics Corporation)

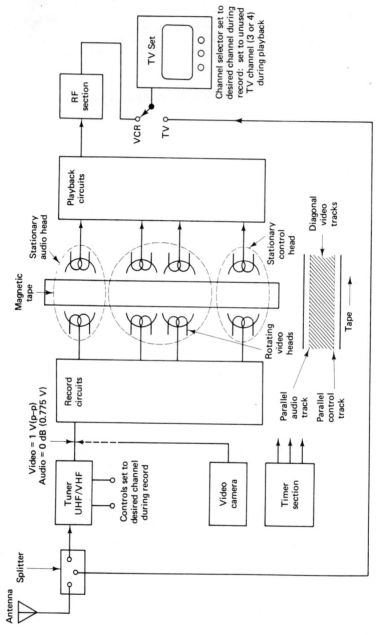

FIGURE 1-2 Basic functional sections of a VCR

found in early-model VCRs use conventional mixer/local-oscillator circuits. The tuners in modern VCRs have some form of frequency synthesis, using *phase-locked loops* (PLLs). Typically, tuner output to the record circuit is 1 V (peak to peak) for video, and 0 dB (0.775 V) for audio, which is generally compatible with most NTSC-type TV sets.

The record circuits function to convert tuner output into electrical signals used by the heads to record the corresponding information on the magnetic tape along tracks. There are three tracks shown (audio or sound, video or picture, and control or synchronization). Note that the audio and control tracks are parallel to the tape, whereas the video track is diagonal. As is explained in the following sections, the video track is recorded diagonally to increase tape writing speed and thus to increase the frequency range necessary to record video signals.

The playback circuits function to convert information recorded on the tracks and picked up by the heads into electrical signals used to modulate the RF section. In the simplest of terms, the RF section is a miniature TV broadcast station operating on an unused TV channel (typically channel 3 or 4, but possibly 5 or 6 in a few rare cases). The output of the RF section is applied to the TV set.

During normal operation, you select the channel you wish to record using the VCR tuning controls. This need not be the channel being watched on the TV set. Similarly, the TV set need not be on while recording with the VCR. You then turn on the timer and the program or programs are recorded. Typically, the VCR records up to about 8 hours, depending on the timer setting and tape speed.

When you are ready to play back the recorded program, you select the appropriate unused TV channel (3 or 4) using the TV set channel controls. Then you turn on the timer and play back the program, using the TV set as a display device or monitor. Some modern VCRs have monitor outputs (audio and video) which can be applied directly to a monitor TV (or another VCR), thus bypassing the RF section.

1-2 RELATIONSHIP OF RECORDING/PLAYBACK HEADS

Recording of the audio and synchronization control signals on a VCR is a relatively simple matter when compared to recording the video signals. The control signals are typically 60 Hz, whereas the audio signals rarely go below about 20 Hz or above 20 kHz. (A typical audio range is 50 Hz to 10 kHz.) The methods used in audio tape recorders are adequate for both of these signals. However, typical video signals go from direct current (0 Hz) up to about 4.2 MHz. There are three methods used to increase the frequency range or writing speed to accommodate the video signals used in VCRs: *frequency mod-*

8   Introduction to Videocassette Recorders

ulation (FM), *micro head gaps,* and *rotating heads* to increase *relative speed* between head and tape.

Figure 1-3 shows the relationship of the recording and playback heads on a typical two-head VCR. The same principles apply to four-head models. Instead of moving the tape at a high speed, the video heads are rotated to produce a high relative speed between head and tape. Actual tape speed is in

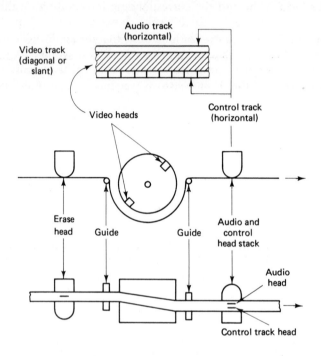

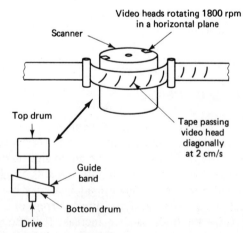

FIGURE 1-3   Relationship of heads and tape movement

the range of 2 cm/s, and the video heads are rotated at 1800 rpm. This results in a relative speed in the range of 5 to 7 m/s (225 to 275 in./s).

Note that the video heads rotate in a horizontal plane (on a *drum, cylinder,* or *scanner,* depending on which literature you read), while the tape passes the heads diagonally. This is known as *helical scan,* and produces *slant tracks* or *diagonal tracks* for video recording. The audio head and control track head (mounted one above the other) are stationary and are separate from the video heads, as is the erase head.

1-3 RELATIONSHIP OF ROTATING HEADS TO VIDEO FIELDS AND FRAMES

Figure 1-4 shows a simplified diagram of the relationship between the video heads and the video tracks recorded on tape. As shown, video heads A and B are positioned 180° apart on the drum or cylinder, which rotates at a rate of 30 times a second (1800 rpm). The tape is wrapped around the drum to form an omega (Ω) shape. The tape then passes diagonally across the drum surface to produce the helical scan. Since there are two heads on the drum (which is rotating at 30 rps), each head contacts the tape once each 1/60 s. Each head completes one rotation in 1/30 s, and one slant track is recorded on the tape during half a rotation (1/60 s).

Since the tape is moving, after the first head has completed one track on the tape, the second head records another track immediately behind the first track, as shown in Fig. 1-4. If head A records during the first 1/60 s,

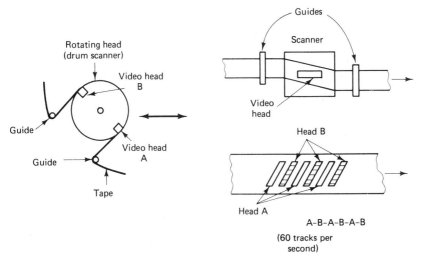

FIGURE 1-4 Simplified diagram of the relationship between the video heads and the video tracks recorded on tape

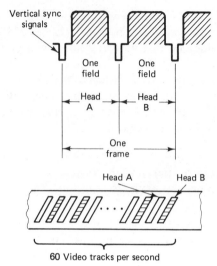

FIGURE 1-5 Theoretical relationship among tracks, fields, frames, and the television vertical sync pulses

head B records during the second 1/60 s. The recording continues in the pattern A-B-A and so on. During playback, the same sequence occurs (the heads trace the tracks recorded on the tape and pick up the signal, producing an FM signal that corresponds to the recorded video signal).

Figure 1-5 shows the theoretical relationship among tracks, fields, frames, and the television vertical sync pulses. Since there are two heads, 60 diagonal tracks are recorded every second. One field of the video signal is recorded as one track on the tape, and two fields (adjacent tracks A and B) make up one frame. In actual practice, there is some overlap between the two tracks. As an example, the video signal recorded by head A (just leaving the tape) is simultaneously applied to head B (just starting its track). During playback, this overlap is eliminated by electronic switching so that the output from the two heads appears as a continuous signal.

1-4 THE BASIC VCR SERVO SYSTEM

It is obvious that no matter how precisely the tracks are recorded, the picture can not be reproduced if these tracks are not accurately traced by the rotating heads during playback. In addition to mechanical precision, both Beta and VHS systems use an automatic self-governing arrangement generally known as the *servo system*. Early-model VCRs generally use a single motor that drives both the head scanner and tape capstan through belts and gears. Modern VCRs generally use separate direct-drive (DD) motors for scanner and capstan. The

DD motors are electronically synchronized as to phase and frequency by the servo system circuits. We discuss modern VCR servo systems in Chapters 7 and 8. For now, we will consider the basic or typical servo system.

Figure 1-6 shows operation of a basic servo system for a typical two-head VCR. As shown, the vertical sync pulses of the TV broadcast signal are used to synchronize the rotating heads with the tape movement. The TV sync pulses are converted to 30-Hz control signals (often referred to as the *CTL signal*). This CTL signal is recorded on the tape by the separate stationary control track head (which is on the same stack as the audio head, as shown in Fig. 1-3).

One major purpose of the servo system in Fig. 1-6 is to rotate the cylinder at precisely 30 Hz (1800 rpm) during record. Note that 30 Hz is one half the vertical sync frequency (60 Hz) of the input video signal. With a 30-Hz speed, the vertical blanking period can be recorded at any desired point on each video track. In television, the vertical blanking occurs at the bottom of the screen, where blanking does not interfere with the picture. For this reason, the vertical sync signal is recorded at the bottom (or start) of each video track. This is shown in Fig. 1-7, which is the typical magnetic tape pattern used in VHS.

In the system of Figs. 1-6 and 1-7, there are two heads (channel 1 and channel 2), and each head traces one track for each field. Two adjacent tracks or fields make up one complete frame. To ensure that there are no nonimages or blanks in the picture, the information recorded on tape overlaps at the changeover point (from one head to another). This changeover point must also occur at the bottom of the screen, where the changeover does not interfere with the picture. For that reason, the vertical sync signal is recorded precisely in a position 6.5H from the changeover time of the channel 1 and channel 2 tracks. (The term "1H" refers to the period of time required to produce one horizontal line on the TV screen, or approximately 63.5 μs, and is sometimes referred to as *line period*. 6.5H is 6.5 times 1H.)

This precise timing requires that the speed and phase of both the cylinder motor and capstan motor be controlled (since the cylinder motor determines the position of the heads at any given instant, while the capstan motor determines the position of the tape). In older VCRs, the synchronization is achieved by driving both capstan and cylinder from a common motor through belts. In a modern VCR servo system, such as shown in Fig. 1-6, five separate (but interrelated) signals are used to get the precise timing. The following paragraphs describe each of these signals and, in general terms, how they are used. The tables of Fig. 1-6 summarize the signal functions. The signals and functions in modern VCRs are described fully in Chapters 7 and 8.

Cylinder FG pulses. The cylinder FG pulses are developed by a generator in the video head cylinder. In the system of Fig. 1-6, the generator consists of an eight-pole magnet installed in the cylinder rotor and a detection

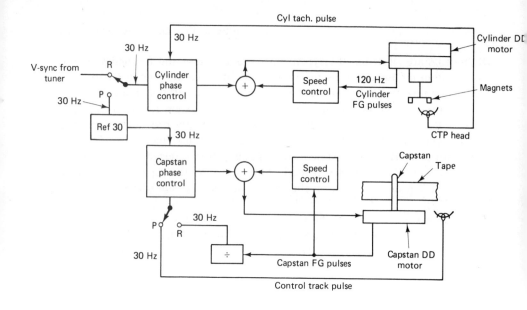

Pulse	Ferquency (Hz)	
Cylinder FG	120	
Capstan FG	SP (2 hour)	720
	LP (4 hour)	360
	EP (6 hour)	240
	Slow (slow motion)	120
	Quick (fast motion)	720
	Search	2160
Cylinder tach	30	
REF 30	30	
Control track	30	

System		Mode	Reference signal	Waveform	Comparison signal	Waveform
Phase	Cylinder	REC	$\frac{1}{2}$ V-sync	Sample pulse	Cylinder tach pulse	Trapezoid
		PB	REF 30			
	Capstan	REC	REF 30	Trapezoid	1/24 Capstan FG	Sample pulse
		PB	REF 30		Control track pulse	
Speed	Cylinder	PB/REC	Cylinder FG	Sample pulse	Cylinder FG	Sawtooth
	Capstan		Capstan FG		Capstan FG	

FIGURE 1-6 Basic servo system for a typical two-head VCR

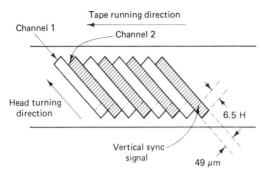

FIGURE 1-7 Typical magnetic tape pattern used in VHS

coil in the stator. When the cylinder rotates at 30 rps, the stator coil detects the moving magnetic fields and produces the cylinder FG pulses at a frequency of 120 Hz.

Capstan FG pulses. The capstan FG pulses are developed by generator in the tape capstan and are applied to the capstan speed control circuits, as well as the capstan phase control circuits (through a divider) during record. In Fig. 1-6, the generator consists of a 240-pole magnet installed in the lower part of the capstan shaft and a detection coil in the stator. (In many modern VCRs, all of the servo control pulses—cylinder, capstan, tach, reference, and control—are developed by Hall-effect generators, as discussed in Chapters 7 and 8.) When the capstan rotates, the stator coil detects the moving magnetic fields and produces the capstan FG pulses. The frequency of the capstan FG pulses depends on the capstan speed (which also controls tape speed). The tables of Fig. 1-6 show some typical capstan FG pulse frequencies for various playing times and tape speeds. Note that not all VCRs have the same six play modes or tape speeds shown in Fig. 1-6.

Cylinder tach pulse. The cylinder tach pulses (CTPs) are developed by another generator in the cylinder, and are applied to the cylinder phase control circuits. In some VCRs the generator consists of a pair of magnets installed symmetrically in a disk in the lower part of the cylinder shaft, and a stationary pickup head. Hall-effect generators are used in other VCRs. With the system of Fig. 1-6, the CTP pickup head detects the moving magnetic fields when the cylinder rotates. The pulse frequency is a constant 30 Hz. In effect, the tach pulse indicates video head channel switching, and is used as a comparison signal in the cylinder phase control circuits during both record and playback.

REF 30 pulse. The reference signal for the phase control system of *both* the capstan and cylinder motors is taken from a crystal oscillator with a frequency of 32.765 kHz. A frequency of 30 Hz is obtained when the crystal oscillator signal is divided. The REF 30 pulse is used for the cylinder phase control circuit only during playback. During record, the cylinder phase control receives broadcast V-sync pulses from the tuner.

Control track pulse. The 30-Hz control track pulses are the broadcast V-sync pulses recorded on tape during record. At playback, the pulses are picked up by the control head and applied to the capstan phase control circuit.

1-5 RELATIONSHIP OF LUMINANCE AND CHROMA SIGNALS

Figure 1-8 shows the typical sequence in recording and playback of the luminance signal on a VCR. During record, the entire luminance signal (from sync tips to white peaks) is amplified and converted to an FM signal that varies in frequency from about 3.5 to 4.8 MHz for Beta systems, or 3.4 to 4.4 MHz for VHS. During playback, the FM signal is demodulated back to a replica of the original luminance signal. Note that this provides an FM luminance bandwidth on tape of about 1.3 MHz for Beta and 1 MHz for VHS.

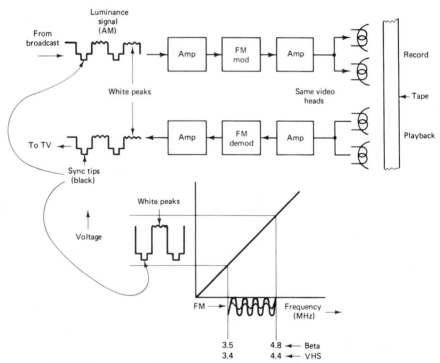

FIGURE 1-8 Typical sequence in recording and playback of the luminance signal on a VCR

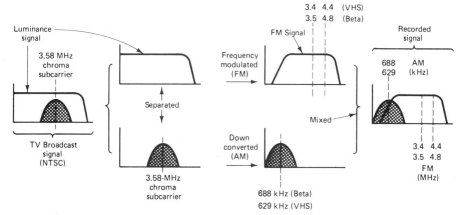

FIGURE 1-9 Typical sequence for down conversion (color under) of a 3.58-MHz chroma subcarrier

Color TV information is transmitted on the 3.58-MHz chroma subcarrier. Color at any point on the TV screen depends on the instantaneous amplitude and phase of the 3.58-MHz signal. In VCRs, the 3.58-MHz subcarrier is *down-converted* to a frequency of 688 kHz for Beta or 629 kHz for VHS, and recorded directly (AM, not FM) on tape. Figure 1-9 shows the typical sequence for such down-conversion (known as a *color-under* system, since the color signal frequency is always well below the luminance signal frequency).

There are several advantages for the color-under system. No bias is needed to record the chroma signal. Since the FM luminance signal is recorded together with the chroma signal (both signals on the same video heads), the FM luminance signal acts as a bias. Electronic stability is good since the color signal (including sidebands of about ±500 kHz) is far removed below the lowest luminance signal frequency.

1-6 BETA SYSTEM BASICS

Note that the discussion in this section applies to Betamax, developed and manufactured by Sony. However, the principles also apply generally to other Beta systems. The basic Beta system is known as *guard bandless* recording or *zero guard band* recording. On broadcast-type VTRs, the tape is driven against the video heads so there is unrecorded vacant space between the video tracks. This blank area, or guard band, is necessary to eliminate *crosstalk* between tracks. Figure 1-10 shows a comparison of tapes with and without guard bands. In the Beta system, the problem of crosstalk is eliminated by *azimuth recording* and *phase inversion* (or PI).

16   Introduction to Videocassette Recorders

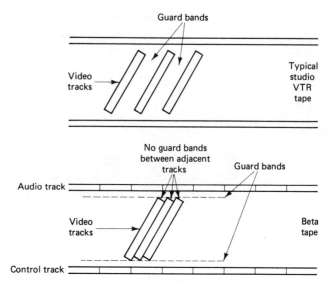

FIGURE 1-10 Comparison of tapes with and without guard bands between adjacent video tracks

1-6.1 Azimuth recording

Figure 1-11 shows how the azimuth recording (or *azimuth-loss recording*) principle is used to minimize crosstalk between adjacent tracks. The two video heads (or pairs of video heads on four-head models) are mounted so that one head is at a different angle from the other. The angle for one head (head A) is +7° from the reference point (typically at right angles to tape movement), whereas head B is -7° from the reference. This produces a 14° difference between the heads and, during playback, a strong signal is picked up only when head A traces over track A. If head B runs over track B for any reason, the track B signal is weak and does not produce interference or crosstalk.

1-6.2 Phase-inversion (PI) color recording

Phase inversion is used to minimize crosstalk (primarily in the chroma channel, which requires more protection). In the simplest of terms, the chroma signal to be recorded on track A is phase-inverted by 180° with every line period (1H), while the chroma signal recorded on track B remains continuously in the same phase. At playback, both track A and track B signals are restored to the same phase relationship.

Figure 1-12 shows how the chroma signal is recorded using phase inversion. The line-by-line phase inversion of the chroma signal is done during the conversion process from 3.58 MHz to 688 kHz and applied to head A. The

Beta System Basics 17

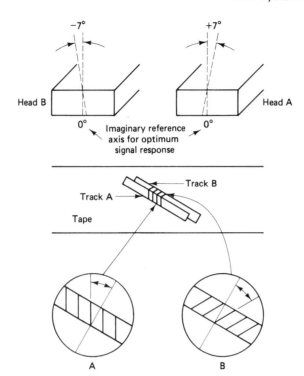

FIGURE 1-11 Displacement of video-head azimuth in Beta systems

3.58-MHz chroma signal is mixed with a 4.27-MHz reference signal (which is phase-inverted each time that head A is in contact with the tape). The resultant 688-kHz signal is amplified and applied to the video heads (head A inverted, head B not inverted).

During playback, the 688-kHz signal is converted back to 3.58 MHz. In an exact counterpart of the recording process, the head A signal is again phase-inverted using the same 4.27-MHz signal used during record. Adding the recovered line signal (head A) to the adjacent line signal (head B) restores the signal to normal. However, the phase of the crosstalk component in the 3.58-MHz playback chroma signal remains phase-inverted at every other line.

The playback chroma signal is then passed through a *comb filter* using a 1H *delay line* and a resistive matrix, as shown in Fig. 1-13. Both the delayed and nondelayed signals are added together in the resistive circuit, with the result that the crosstalk component is cancelled out and the normal chroma signal component is double in amplitude.

In addition to canceling crosstalk, the PI color recording system also minimizes the effect of mechanical jitter or flutter in the scanner servo. Such jitter causes a phase shift and results in poor picture quality. Jitter effects are

18 Introduction to Videocassette Recorders

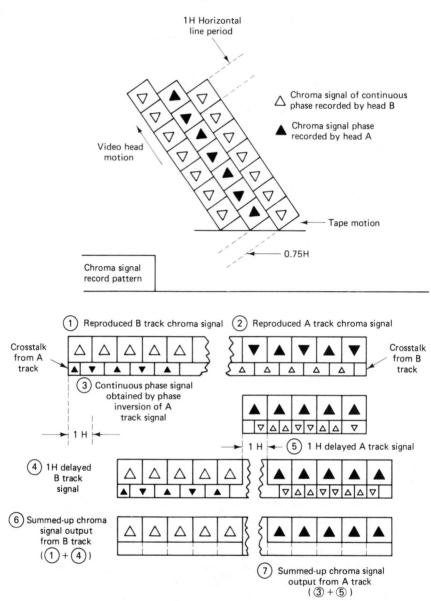

FIGURE 1-12 Chroma signal record pattern and summary of how crosstalk can be eliminated from playback chroma signal

eliminated by locking the frequency and phase of the 4.27-MHz reference signal to the TV horizontal synchronizing (H-sync) signal (known as fH in Beta) during record. At playback, the 4.27-MHz reference signal is locked to the recorded H-sync signal (known as fH*). If there is any jitter component (from

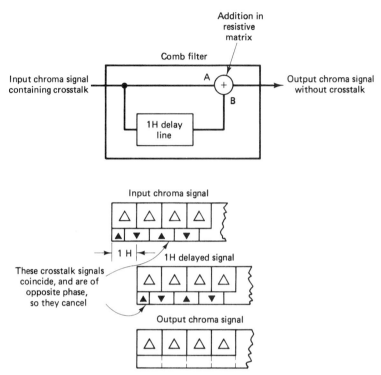

FIGURE 1-13 Function of the comb filter in canceling out crosstalk

any cause, mechanical or electrical), the 4.27-MHz reference is also locked to the jitter, eliminating the jitter effect. This feature is similar to locking the scanner speed to the TV vertical sync during record, and to the recorded V-sync during playback, as described in Sec. 1-4. However, operation of the PI color recording circuits is far more complex.

1-6.3 Basic Beta PI color recording circuit

As shown in Fig. 1-14, there are two phase-locked loops, or PLLs, involved in Beta PI color recording. One PLL is known as the AFC (automatic frequency control) loop, and produces a signal at a frequency 44 times fH, or about 693 kHz. The AFC loop receives the fH input from the TV video signal during record, and an input fH* from the recorded video signal during playback. The AFC loop output (either 44fH or 44fH*) is combined with the output of a 3.57-MHz crystal oscillator in frequency converter 2.

The 3.57-MHz is free-running during record, but is locked in phase to the chroma 3.58-MHz signal during playback. (Note that 3.57 MHz is equal to 3.58 MHz, less ¼fH.) Phase lock of the 3.57-MHz oscillator during playback is done by the APC (automatic phase control) loop. In either playback

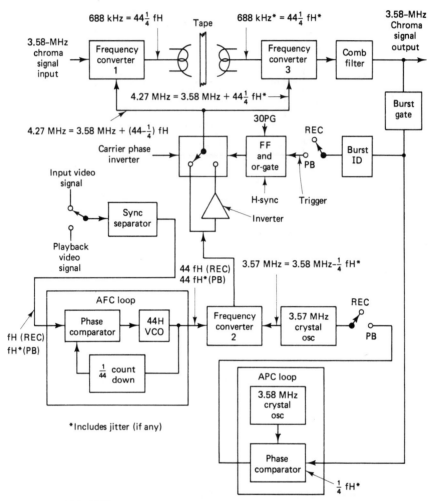

FIGURE 1-14 Basic Beta PI control recording circuit

or record, the 3.57-MHz signal is combined with the 44fH signal to produce a 4.27-MHz signal which, in turn, is applied to frequency converters 1 and 3 through a *carrier-phase inverter*.

During both playback and record, the 4.27-MHz signal is passed by the carrier-phase inverter (usually a center-tapped transformer). The phase inverter is operated by an FF and OR gate that receives both cylinder tach pulses (called 30 PG pulses in Beta) and H-sync pulses. Both signals are required since the carrier-phase inverter serves to phase-invert the 4.27-MHz with H-sync only when track A is being made. The 30PG pulse overrides the H-sync pulses when track B is being traced by head B.

During record, the 3.58-MHz chroma signal to be recorded is applied to frequency converter 1, where the 4.27 signal is added, resulting in a difference frequency of 688 kHz. Since the 4.27-MHz signal is locked to the H-sync signal, the 688-kHz signal to be recorded is also locked to H-sync.

During playback, the 688-kHz chroma signal from the head is applied to frequency converter 3, where the 4.27-MHz signal is again added, resulting in a difference frequency of 3.58 MHz. The 3.58-MHz chroma output signal is compared with the APC loop oscillator (also 3.58 MHz). Any phase variations due to jitter are used to shift the 3.57-MHz oscillator signal (free-running during record). Since the phase of the 4.27-MHz signal is controlled by the 3.57-MHz oscillator, any phase shift in the 3.58-MHz chroma output signal is eliminated.

Even with the AFC circuit, there is still a possibility that the 3.58-MHz chroma playback signal burst will lock up on a wrong phase of the 4.27-MHz signal (locked in, but 180° out of phase). This condition is prevented by the *burst ID* (burst identification) circuit that compares the phase of the 4.27-MHz reference signal with the 3.58-MHz chroma signal during playback. If the APC system has locked up on the wrong phase, the carrier-phase inverter FF circuit is switched by a trigger pulse developed in the burst ID circuit. The burst ID compares the phase of the 3.58-MHz chroma playback signal for each horizontal line and produces the corrective pulse whenever the phase-invert FF switch has locked on the incorrect phase.

1–6.4 Typical overall functions of a Beta VCR during record

Figure 1–15 is a typical overall block diagram of a Beta VCR during record. Note that the circuits shown are very general in overall purpose, and therefore apply to both early-model as well as modern Beta VCRs.

Tuner (VIF/SIF detection). The TV broadcast channel signal is converted to a 45.75-MHz video IF (VIF) and 41.25-MHz sound IF (SIF). The NTSC composite video signal is extracted by a video detector. The audio signal is converted to 4.5 MHz by an audio IF second detector. The low-frequency audio signal is extracted by means of FM detection.

Luminance signal (FM). With AGC maintaining constant NTSC composite video signal levels, the luminance and chroma signals are separated. Then, to convert the wide-frequency band (0 to 4.2 MHz) luminance signal into an easily recordable FM signal, a clamp circuit (among many other circuits) matches up the level of the luminance signal, which is then processed to reduce crosstalk and improve the signal-to-noise ratio. An FM modulator converts the luminance signal to a 3.5- to 4.8-MHz FM signal.

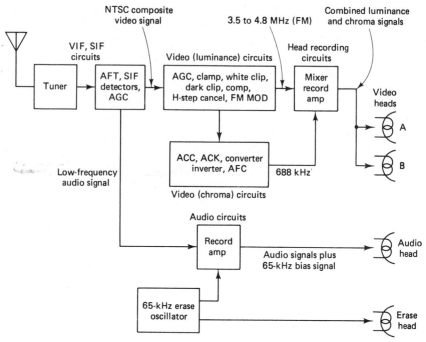

FIGURE 1-15 Typical overall block diagram of a Beta VCR during record

Chroma signal. The separated chroma signal (the 3.58-MHz subcarrier, or so-called "color burst") is mixed with a 4.27-MHz reference signal and converted to a frequency of 688 kHz. Azimuth recording and phase inversion are used to eliminate the problem of crosstalk between tracks.

Head recording amplifier. The FM-modulated signal (3.5 to 4.8 MHz) and the chroma signal (688 kHz) are combined and amplified for optimum recording performance before application to the video heads A and B.

Audio recording. Because of the nonlinear response of magnetic tape, preemphasis is applied during record and deemphasis is applied during playback of the audio signal. A 65-kHz bias signal is superimposed on the audio signal during record to raise efficiency and reduce distortion. The same 65-kHz oscillator output is also used to erase previously recorded signals on the magnetic tape.

Typical Beta video signal flow during record. As shown in Fig. 1-16, the AGC section maintains a constant output level regardless of picture

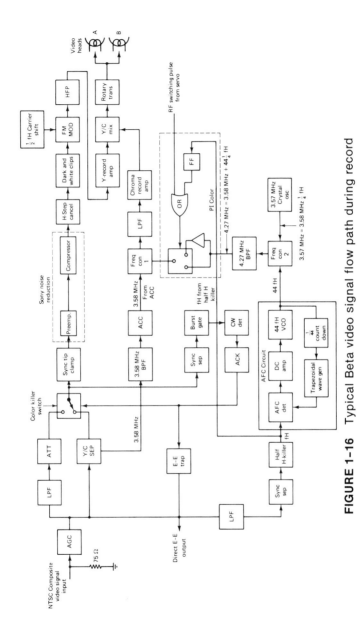

FIGURE 1-16 Typical Beta video signal flow path during record

brightness. Generally, the familiar TV sync type of AGC is used. The LPF (low pass filter) removes the sound IF signal (4.5 MHz) and other unnecessary high-frequency components. The ATT (attenuator) reduces the signal amplitude to a suitable level. The Y/C separator separates the luminance (Y) and chroma (C) components of the color input. The sync tip clamp lines up the sync signal level. The dc component of the video signal is normally removed by capacitive coupling in the signal path. Clamping is used to match up the sync signal since the sync signal tip becomes the low-frequency (3.5-MHz) reference for the FM signal (as shown in Fig. 1–8).

The E-E trap ensures that the direct E-E output (for a monitor TV) switches to the black-and-white mode when signal conditions are bad. Under such bad conditions, the chroma signal level of the input video signal is low, and the VCR color killer (ACK, or automatic color killer) circuit operates, switching the recording mode to black and white.

The term *E-E*, or electric-to-electric, can be explained as follows. During record, the VCR record output circuit is connected to the playback input so that the video signal to be recorded can be monitored on a TV, if desired. Since the magnetic components (heads, tape, etc.) have nothing to do with this signal (the signal is passed directly from one electrical circuit to another), the function is called the E-E mode. When the heads and tape are involved in the normal record/playback cycle, the term *V-V*, or video-to-video, is sometimes used.

The Sony noise reduction system includes both preemphasis and compressor functions to produce a nonlinear emphasis of the video signal during record. During playback, deemphasis (having the opposite characteristics of the emphasis applied during record) is used to reproduce the original signal.

The H step cancel circuit corrects nonuniformity in H-sync spacing. In areas where reception conditions are bad, the H-sync signal may not be uniform, resulting in possible crosstalk problems during playback. The H step cancel circuit maintains constant H-sync spacing.

The dark and white clip circuit clips unwanted high-level pulse components arising from preemphasis to help stabilize the picture. Dark clipping cuts off excess sync signal excursions, while white clipping prevents whiter-than-white level (overmodulation).

The HPF (high-pass filter) differentiates the FM modulator output and passes the differentiated peaks. Differential recording is used because of undesirable self-demagnetization by high-frequency components when the luminance signal (FM) is recorded.

The ½fH carrier shift circuit produces a ½fH difference between the carrier frequencies of the field A period (during the A head period) and field B period. Without such a circuit, there is a possibility of "beating" due to crosstalk between adjacent tracks during playback. The ½fH carrier shift

Beta System Basics 25

circuit shifts the beat frequency by a ½-offset relationship, making the beat unnoticeable on the TV picture.

1-6.5 Typical overall functions of a Beta VCR during playback

Figure 1-17 is a typical overall block diagram of a Beta VCR during playback. Again, the circuits are very generalized and apply to both early-model and modern Beta VCRs.

FM playback amplifier. The FM luminance signal and the 688-kHz chroma signal picked up by the video heads are amplified by a high-S/N-ratio amplifier, and frequency response is corrected. Then the playback outputs are mixed to form a single continuous signal.

Luminance signal playback. A high-pass filter is used to separate the FM component of the playback signal. A dropout compensation circuit (DOC) operates if dropout is present. When there are dirt particles or scratches on

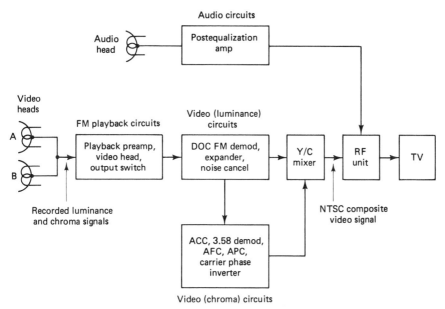

FIGURE 1-17 Typical overall block diagram of Beta VCR during playback

the tape, the video signal may not be picked up by the heads, and noise can appear on the screen. These variations in signal level during playback result in a data-reduction error called *dropout*. After passing the DOC circuit, the luminance signal is frequency corrected and passed through a noise-canceler circuit before emerging as a proper video signal.

Chroma signal playback. The 688-kHz chroma subcarrier is separated from the playback signal by a low-pass filter, and is mixed with an internally generated 4.27-MHz signal to produce a 3.58-MHz chroma subcarrier. At the same time, the comb filter is used to eliminate crosstalk between the video tracks. As discussed in Sec. 1-6.3, the 4.27-MHz signal is processed by the APC and AFC circuits to remove time-base errors (jitter) that may appear during record and playback.

Y/C mixing. The luminance (Y) and chroma (C) signals are mixed to form an NTSC composite color video signal.

Audio playback. The low-frequency signal is picked up by the stationary audio head, undergoes postequalization to correct frequency response and S/N ratio, and is then amplified to a suitable level.

RF unit. The NTSC composite color video signal and the audio signal are converted to a TV broadcast frequency (channel 3 or 4) so that the VCR output signal can be viewed on a conventional color TV set.

Typical Beta video signal flow during playback. As shown in Fig. 1-18, the head preamps amplify the very low amplitude video signals at the heads. The equalizer corrects the frequency response of the head signals and adjusts channel balance so that both head signal levels are equal.

The switcher/mixer combines the signal of both channels and removes any overlap to provide a composite signal output. Since the videotape typically wraps the drum more than 180°, pulses from the servo are used to remove the overlap and produce a continuous output signal, as shown in Fig. 1-19.

The ATT (attenuator) matches the level of black-and-white reproduction with that of color reproduction. The HPF (high-pass filter) passes only the luminance (FM) component and rejects the 688-kHz chroma signal. In the absence of a color signal during playback, the ACK circuit switches the video signal through the HPF to remove any chroma signal or noise that might interfere with black-and-white reproduction.

The DOC senses any dropout and compensates by using the preceding horizontal line (1H) signal. The output of the LPF is applied to the chroma playback circuit, consisting of the frequency converter, comb filter, AFC, APC, PI, and burst ID circuits, as discussed in Sec. 1-6.3.

Output of the DOC is applied to the limiter, which limits amplitude of the FM signal and removes amplitude fluctuation from the playback signal. The FM demodulator/LPF combination changes the FM signal back into the original video signal. Most video FM playback demodulators use a multivibrator/multiplier circuit rather than the familiar balanced FM discriminators. No matter what form of video FM demodulator circuit is used, the circuit is part of an IC and is neither accessible nor adjustable (which also applies to most signal circuits in modern VCRs).

The demodulated video signal is applied through a noise reduction system similar to that described for record (Sec. 1-6.4). Since nonlinear emphasis is applied during record, the opposite process takes place during playback.

The ½fH carrier shift return circuit restores the ½fH carrier shift produced during record. If this carrier shift is not restored, the dc component fluctuates when the FM signal is demodulated.

In the noise canceler, the noise component (high frequency) of the video signal is removed by a high-pass filter. The phase is then inverted and, with reverse phase and the same amplitude, the processed video signal is added back to the original video signal, thus canceling the noise.

During rewind, fast forward, and playback servo startup time, a muting signal from the system control prevents the video signal from being applied to the RF modulator (or to the video output connector).

1-6.6 Automatic tape loading for the Beta system

Figure 1-20 shows the basic auto-loading system used in Beta VCRs. The Beta system uses a so-called U loading or threading system, since the tape appears to form the letter U when fully threaded. When the cassette is inserted into the cassette box, a loading ring picks up the tape, as shown, and then threads the tape around the tape drum (in about three seconds). When the eject button is pressed, the loading ring turns in the reverse direction and the excess tape is taken up by the take-up reel inside the cassette. When the loading ring returns to the original position, and the tape is all back inside the cassette, the cassette automatically rises and is ejected.

Since video cassettes form a parallel two-reel system, the cassette can be easily removed, even if stopped in the middle during record or playback. Since loading takes place automatically as soon as the cassette is inserted, VCRs are no more complicated to operate than an audio tape recorder. Also, the tape cannot be damaged, since no excess pressure is applied and it is not touched by hand.

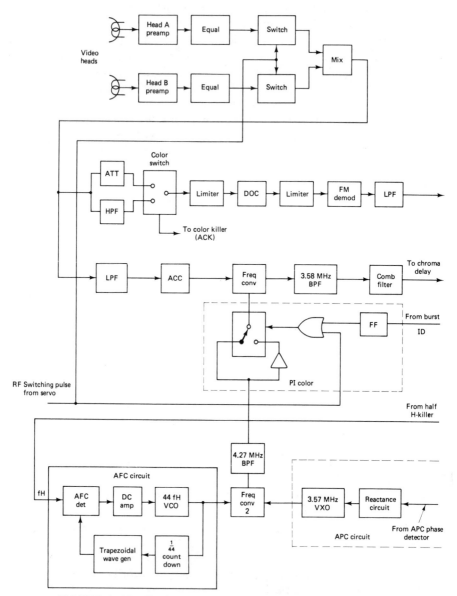

FIGURE 1-18 Typical Beta video signal flow during playback

Beta System Basics 29

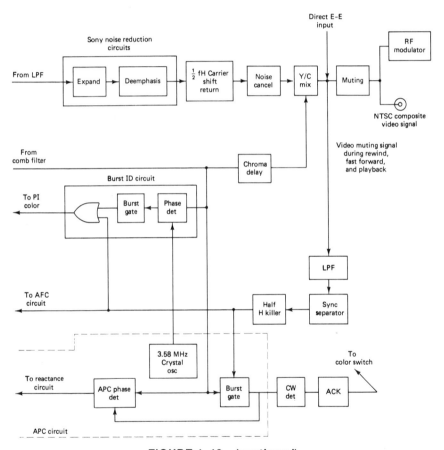

FIGURE 1-18 (continued)

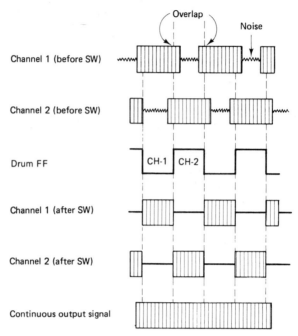

FIGURE 1-19 Switching and mixing process to produce a continuous signal from the video heads

1-7 VHS SYSTEM BASICS

The VHS system also uses high-density recording to get the maximum amount of program information on a given amount of tape. This involves *zero guard band recording* and results in the crosstalk problem described in Sec. 1-6 for Beta. The VHS system also uses *azimuth recording* and *phase inversion* to minimize the effects of crosstalk. The azimuth recording used for VHS is similar to that for Beta. However, VHS uses a ±6° azimuth difference (resulting in a 12° difference between heads) rather than the ±7° for Beta. Also, VHS records the chroma or color information at 629 kHz rather than the 688 kHz for Beta. The 629 kHz is produced by mixing the incoming 3.58-MHz chroma signal with a 4.2-MHz reference, which is phase-inverted and locked to the incoming H-sync signal. Note that 629 kHz is 40 times the H-sync frequency of 15,750 Hz (actually 15,734.26 Hz during a color broadcast).

The phase-inversion system used in VHS is entirely different from that used in Beta. In the basic VHS system, the phase of the 629-kHz color signal being recorded on head A is advanced in phase in increments of 90° at each successive horizontal line. At the end of four lines, the 629-kHz signal is back to original phase. For example, lines 1, 2, 3, and 4 are shifted 0°, +90°,

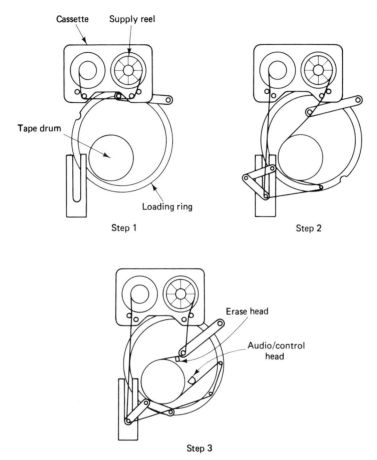

FIGURE 1-20 Basic auto-loading system for a Betamax VCR

+180°, and 270° in succession. When head B is recording, the 629-kHz color signal is shifted in phase (retarded) in the opposite direction (0°, 270°, 180°, 90°). This results in the following pattern:

Line	1	2	3	4	5	6
Head A	0°	90°	180°	270°	0°	90°
Head B	0°	270°	180°	90°	0°	270°

Thus, recorded phase shifts for odd-numbered lines (1, 3, 5) are the same, but are opposite for even-numbered lines (2, 4, 6).

When the 629-kHz color signal is played back, the 4.2-MHz signal is again phase-inverted and mixed with the 629-kHz signal to restore the 3.58-MHz chroma signal. When both the playback 629-kHz and reference 4.2-kHz signals are phase-shifted in the same direction, the effect in the mixer is to restore the 3.58-MHz signal to normal phase. When the playback 629-kHz and reference 4.2-MHz signals are shifted in opposite directions, the phase of the 3.58-MHz chroma signal is shifted on every other line. As discussed in Sec. 1-5.2, when such a signal is phased through a 1H delay line, the crosstalk component is cancelled out and the normal chroma signal component is double in amplitude. This effect is shown in Fig. 1-21.

1-7.1 Typical luminance circuit operation of a VHS VCR during record

As shown in Fig. 1-22, the video signal from the tuner is fed to LPF 2F4, where the 3.58-MHz color signal is attenuated and the video fed to the AGC circuit, which keeps the output level constant. The video signal is amplified and fed to LPF 2F5, which serves as a 3.58-MHz trap to remove the color signal. A pure video signal is amplified by IC2F2 and is fed to the nonlinear emphasis circuit, which emphasizes the luminance signal frequencies by dif-

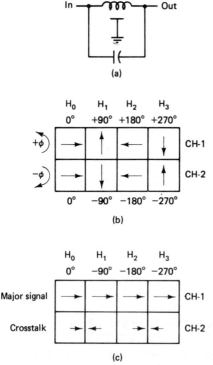

FIGURE 1-21 1H delay functions

VHS System Basics   33

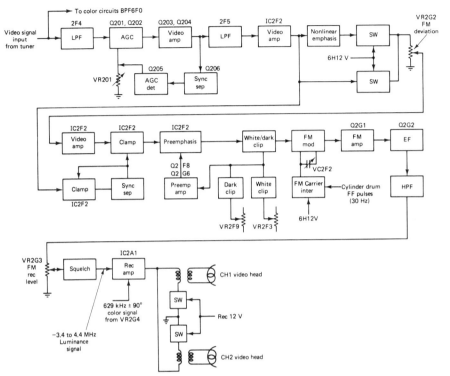

FIGURE 1-22 Luminance (Y) signal flow in a VHS system during record

ferent amounts, depending on playing time (two or six hours). The selection of frequency emphasis is made by a pulse 6H12V applied to the switches.

The nonlinear emphasis circuit output is applied through VR2G2, which sets the FM modulation level. Note that the nonlinear emphasis network is completely bypassed on the two-hour playing mode. In all modes, the signal from the nonlinear emphasis circuit is amplified by the video amplifier (part of IC2F2) and is fed to a clamp where the dc voltage of the video sync tip remains constant regardless of the function in the video signal. This keeps the sync tip at 3.4 MHz (as shown in Fig. 1-8).

The clamped signal is fed to the preemphasis network, where the high-frequency spectrum is emphasized to improve the S/N ratio. The preemphasis of high frequencies is necessary to reduce noise. In FM, the higher the modulated frequency, the more liable modulation is to be influenced by noise. Overmodulation can cause a reverse or negative picture as well as a poor S/N ratio. The white and dark clip circuits are included to prevent overshoot above a specified level.

The white and dark clip circuits are set to the correct level by VR2F3 and VR2F9, respectively. Output from the white and dark clip circuits is applied to the FM modulator, which operates at 3.4 MHz for the sync tips and 4.4 MHz for white peaks. The FM modulator also receives 30-Hz pulses from the FM carrier interleaving circuits. These circuits are similar to the ½fH carrier shift circuits described for Beta, and advance the video signal phase by ½fH for one head or track.

The FM luminance signal is amplified and passed through an emitter follower and high-pass filter, which attenuates the lower end of the FM so as not to interfere with the 629-kHz chroma signal added later in the signal path. The HPF output is applied to a squelch circuit through VR2G3, which sets the FM modulation level. The squelch circuit serves to prevent the signal from being fed to the record amplifier for about 1.5 s after completion of cassette loading. This prevents the recorded signal from being erased if the tape runs near the drum in the middle of loading. The signal passing through the squelch circuit is amplified and applied to the video heads through a rotary transformer. Note that the record amplifier also receives the 629-kHz chroma signal from the color recording system, as described in Sec. 1-6.3.

1-7.2 Typical luminance circuit operation of a VHS VCR during playback

As shown in Fig. 1-23, the reproduced signal from the video heads is applied to the preamps through rotary transformers. Switch circuits process the signals from the two channels to remove any overlap (Fig. 1-19). The composite signal is amplified by IC2A0 and applied to the video amplifier through VR2F4. The reproduced signal is also made available to the color circuits (Sec. 1-6.4) from VR2F4. After being amplified, the signal is passed through 2F0 to extract only the luminance (Y) FM signal and applied to a DOC circuit (detector, gate, 1H delay, mixer). The DOC prevents picture deterioration by supplying a 1H preceding signal if the FM signal is partially missed (dropout) for any reason (dirt, flaws in tape, etc.). The DOC also provides a pulse to the AFC circuits (Sec. 1-7.6).

The FM signal is then fed to a double limiter circuit (high-pass filter, first limiter, low-pass filter, mixer, second limiter) which removes the AM from the FM signal. The luminance signal is then fed to the FM demodulation circuit (which typically uses a delay-line type of phase detection). The demodulated signal is amplified and impedance matched by Q2F4, and only the video signal is taken from 2F2. The video signal is deemphasized (in reverse to the preemphasis at recording, Sec. 1-7.1) by Q2F5. An edge-noise canceler removes the noise, and the video signal is applied to a compensator circuit through Q2F6.

During the six-hour playing mode, the compensator (together with the nonlinear deemphasis and feedback amplifier) returns the nonlinear emphasis supplied during record (Sec. 1-7.1). Note that the nonlinear deemphasis network is turned on during the six-hour mode by the 6H12V signal. The feed-

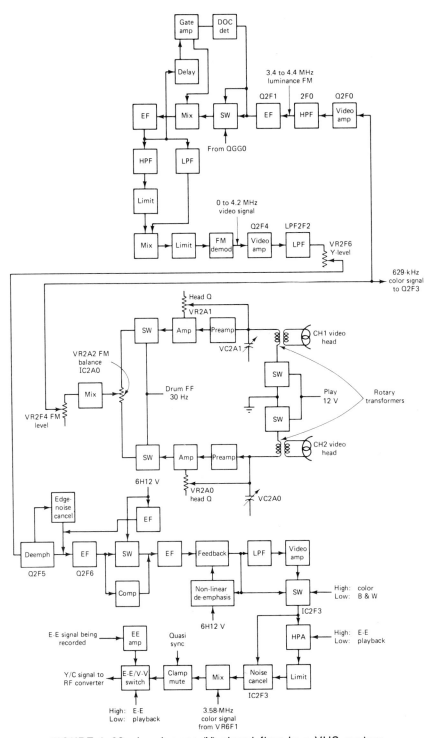

FIGURE 1-23 Luminance (Y) signal flow in a VHS system during playback

back amplifier output is applied to a video amplifier through a low-pass filter (during a color broadcast). For black and white, the low-pass filter is bypassed by action of the color/black-and-white switch in IC2F3. The low-pass filter removes any noise that may arise where the video signal overlaps the demodulated chroma signal.

The signal transmitted through IC2F3 is sent to the noise-canceler circuit of IC2F3, which suppresses any noise in the video signal. The luminance (Y) and chroma (C) signals are combined in the Y/C mixer and then applied to the E-E/V-V switch through the clamp mute circuit. The E-E/V-V switch output is applied to the RF unit, which produces NTSC signals on channel 3 or 4. The E-E amplifier and E-E/V-V switch combination permits a signal being recorded to be monitored on a TV set, if desired (Sec. 1-6.4).

1-7.3 Typical color circuit operation of a VHS VCR during record

As shown in Fig. 1-24, the video from the tuner is passed through Q6F0 to remove only the chroma (3.58 MHz ±500 Hz), and is amplified. The color signal is then fed to Q6F2 to boost the burst signal by 6 dB. The burst is fed to the automatic color control (ACC) through impedance-matching Q6F1.

The signal at pin 11 of IC6F0 is applied to the color control detector circuit through Q6F3 and a burst gate circuit. The color signal peak is detected and produces a control voltage which is applied to the ACC circuit. This detected voltage controls the ACC to maintain the color signal at a certain voltage level. The color signal is then applied to the main converter, and mixed with a reference signal to produce the desired 629 kHz for recording on tape (with the luminance signal, Sec. 1-7.1).

The reference signal applied to the main converter is developed by mixing a 3.58-MHz signal from the VXO (variable crystal oscillator) and a 629-kHz ±90° signal from the AFC circuit (Sec. 1-7.6). Note that the term "±90°" applied to a signal in VHS means that the signal is rotated or shifted in phase every 1H period. The 3.58-MHz VXO and 629-kHz signals are combined in the subconverter to produce a 4.2-MHz signal. This 4.2-MHz signal is passed through a bandpass filter to the main converter, where the signal is combined with the 3.58-MHz chroma signal to produce a 629-kHz ±90° signal. The resultant signal is then fed to the record amplifier (with the luminance signal) through a low-pass filter, emitter follower, and color record level control VR2G4.

1-7.4 Typical color circuit operation of a VHS VCR during playback

As shown in Fig. 1-25, the reproduced signal from IC2F0, applied through the FM level adjust VR2F4 (Sec. 1-6.2), is amplified by Q2F3. At this point the signal contains both luminance and chroma. Only the 629-kHz ±90°

VHS System Basics   37

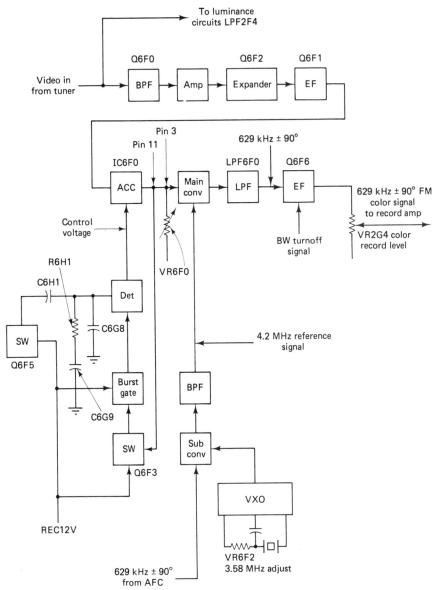

FIGURE 1-24 Chroma (C) signal flow in a VHS system during record

chroma signal is passed by LPF2F0, amplified by Q6F0, and applied to the ACC circuit through Q6F1.

The chroma signal is maintained at a constant level by the ACC circuit, and is mixed with a 4.2-MHz ±90° signal from the APC circuit (Sec. 1-7.5)

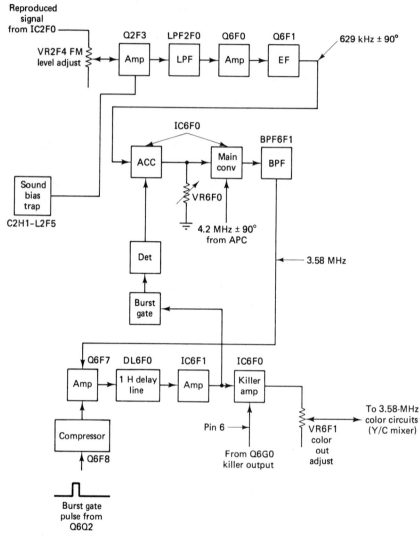

FIGURE 1-25 Chroma (C) signal flow in a VHS system during playback

in the main converter. The resultant 3.58-MHz signal is amplified by Q6F7 after being passed by BPF6F1.

The compressor circuit operates whenever there is a burst gate pulse to reduce gain in amplifier Q6F7. This is necessary to restore the burst signal to a normal level at playback. As discussed in Sec. 1-7.3, the burst signal is increased in amplitude by about 6 dB during record.

The restored chroma signal is passed thrugh a 1H delay line to remove crosstalk. The 1H delay line output is amplified by IC6F1 and is fed to the killer amplifier of IC6F0. This killer amplifier is a switch circuit that allows the signal to pass only when there is a color signal carrier present in the video signal. When the playback is black and white (no color carrier), the color killer prevents a color signal from being applied to color-out control VR6F1. This function eliminates noise components from the chroma circuit being applied during a black-and-white playback. The color killer is operated by signals from Q6G0. When pin 6 of IC6F0 is high, the color signal passes through VR6F1 and is superimposed on the luminance (Y) signal (Sec. 1-7.2).

1-7.5 Typical APC circuit operation of a VHS VCR

As shown in Fig. 1-26, VXO IC6F2 operates as a fixed 3.58-MHz oscillator *during record*. The output of IC6F2 is mixed with the 629-kHz $\pm 90°$ signal from the AFC (Sec. 1-7.6) in the subconverter to form a 4.2-MHz reference voltage, as discussed in Sec. 1-7.3. The VXO output is also applied to the phase detector killer IC6F2 through diode switches D6G6 and D6G7.

During playback, VXO IC6F2 operates as a phase-locked 3.58-MHz oscillator. The phase of IC6F2 is controlled by an error voltage from the phase detector APC IC6F2. This phase detector receives and compares two 3.58-MHz inputs. One input is the playback color burst (which includes any phase shifts due to jitter), while the other input is from a fixed 3.58-MHz oscillator IC6F5. If there are any phase differences between the two signals, the error voltage produced by the phase detector shifts the phase of the VXO to correct the condition. Since the playback color burst phase is controlled by the VXO, any phase shift in the playback 3.58-MHz color signal is eliminated.

Note that the term $\Delta\phi$ (delta phi), applied to a signal, means that the signal has been shifted in phase or is of differing phase. In the case of the 3.58-MHz signal applied to the phase detector in VHS, the term means that the signal contains any possible jitter effect which could shift the phase.

The killer circuits of Fig. 1-26 have two functions. First, they prevent a color signal from being passed when the signal is black and white only (to eliminate color circuit noise from being mixed with the black-and-white signal). Second, the killer circuits provide an identification pulse (ID) which is used in the AFC circuit to prevent 180° out-of-phase lockup. This is similar to the burst ID pulse used for Beta as described in Sec. 1-6.3.

During color operation the 3.58-MHz color burst signal is passed through the burst gating circuit Q6F9 to the phase detector killer IC6F2, which also receives a 3.58-MHz signal from either the 3.58-MHz oscillator IC5F5 (during playback) or the VXO (during record). The two signals are compared in phase by the phase detector killer. If both signals are of the same phase, the output of the killer detector IC6F3 becomes low and the killer output Q6G0 goes high. This high is applied to pin 6 of IC6F0. As discussed in Sec. 1-7.4, with a high at pin 6 of IC6F0, the color signal is passed.

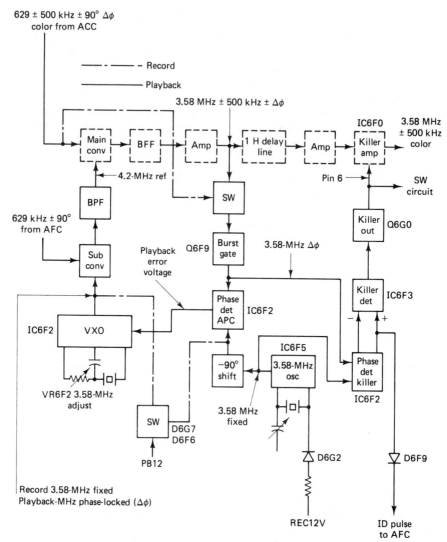

FIGURE 1-26 Typical APC circuit operation

During black-and-white operation there is no 3.58-MHz color burst signal, so the phase detector killer IC6F2 sees only one signal. The output of IC6F3 then goes high, and the output of Q6G0 goes low. This low cuts off the killer amplifier and prevents passage of color signals (or color noise).

Also during color operation, if the 3.58-MHz color burst is exactly 180° out of phase with the IC6F5 3.58-MHz oscillator (locked in phase, but 180° out), the phase detector killer IC6F2 develops a burst ID pulse, which is applied through D6F9 to the AFC (Sec. 1-7.6).

1-7.6 Typical AFC circuit operation of a VHS VCR

Figure 1-27 is a block diagram of typical AFC circuits. Note that the AFC system operates in much the same way for both record and playback. However, during record, the AFC uses H-sync pulses in the video signal from the tuner. During playback, the AFC uses the H-sync signals recorded on tape.

The AFC system has five inputs and one output. The five inputs include the video H-sync pulses, a dropout pulse from the DOC (Sec. 1-7.2), a 30-Hz cylinder signal from the servo (Sec. 1-4), a color burst ID pulse from the APC (Sec. 1-7.5), and a 3.58-MHz fixed or phase-corrected signal from the VXO in the APC circuits. The output of the AFC circuit is a 4.2-MHz $\pm 90°$ signal (fixed reference during record or phase-corrected during playback).

The video signal (from tuner or playback) is applied to a sync separator where only the vertical and horizontal sync signals (V-sync and H-sync) are passed. The resultant signal is then applied to the HSS (H-sync separator) where only H-sync is passed. The H-sync signals (or fH, as they are referred to in most VCR literature) are shaped into a 2-μs pulse by a HD (horizontal drive) circuit. The output from the HD circuit is adjusted to exactly 2 μs by VR6F3, and is applied to an AND gate. The other AND gate input is normally high so that the 2-μs fH pulses can pass. However, if there is a dropout (Sec. 1-7.2), the other AND gate input goes low, preventing the fH pulses from passing.

The output of the AND gate is applied to an AFC circuit within IC6F5. This AFC circuit also receives an fH' (fH prime) signal developed by a 2.5-MHz VCO. Note that the actual frequency of the 2.5-MHz oscillator is 160 times the H-sync frequency of 15,750 Hz (for black and white), or 160 times 15,734.26 Hz (for color). Note that the term *prime* applied to a signal here means that the signal has been locked in frequency to some other signal (to the H-sync signals in this case).

The 160fH' from the VCO is divided by four into 40fH' through operation of a 1/490° shift and switch circuit which is operated by a 4-bit counter. The 40fH' output of this circuit is further divided by 10 to produce 4fH', and by one fourth to produce 1fH' (or simply fH'). This fH' is fed back to the AFC. If there is any difference in frequency between the fH signal coming from the AND gate and the fH' signal originating at the VCO, the VCO is shifted in frequency by an error-correction voltage developed in the AFC circuit, locking the VCO precisely onto the H-sync frequency.

The 4-bit counter (operated by the fH' and cylinder pulses) produces switch signals which select each of four signals from the 1/490° shift circuit, in sequence. Each of the four signals is shifted by 90° from the previous signal, as shown. In effect, the switch can be thought of as a rotary switch where the rotor direction and speed are determined by the 4-bit counter. The counter supplies the pulses to the switch each time an H-sync pulse is applied.

As shown in Fig. 1-28, the channel 1 signals are advanced in phase by

42 Introduction to Videocassette Recorders

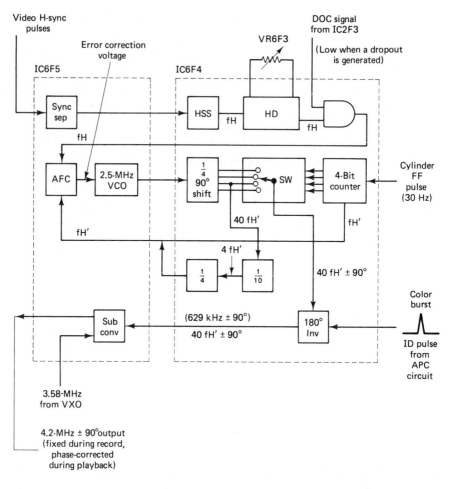

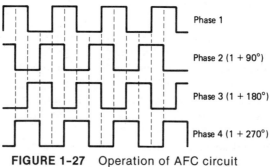

FIGURE 1-27 Operation of AFC circuit

VHS System Basics   43

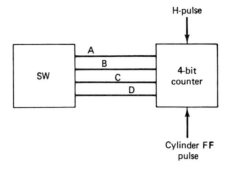

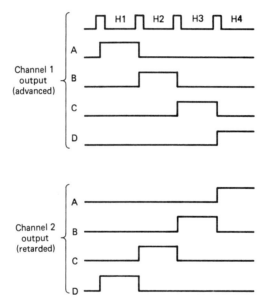

FIGURE 1-28 Operation of a 4-bit counter to advance and retard signals by 90°

90°, whereas the channel 2 signals are retarded or delayed in phase by 90°. These signals (40fH' ±90°) are applied through the 180° inverter circuit to be mixed in the subconverter with the 3.58-MHz signal from the VXO, and result in a 4.2-MHz ±90° that is precisely locked to the VCO.

The 180° is operated by the burst ID pulse from the APC. As discussed in Sec. 1-7.5, the ID pulse occurs only when the 3.58-MHz color burst is 180° out of phase with the 3.58-MHz oscillator (locked in phase, but 180° out). The inverter normally passes the 40fH' ±90° signal without change. However, if the burst ID pulse is present (indicating an undesired 180° lockup), the inverter reverses the phase of the 40fH' ±90° signal to correct the condition.

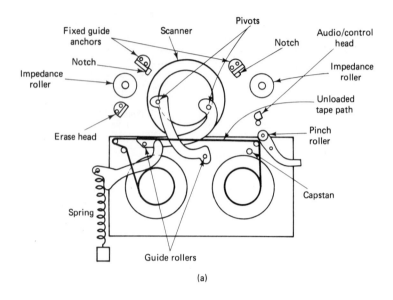

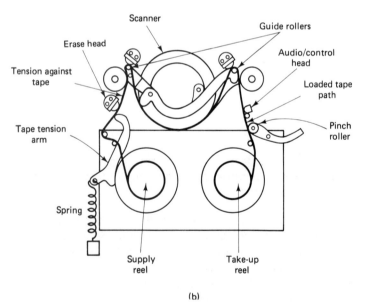

FIGURE 1-29 Typical tape loading system for early-model VHS VCRs

1-7.7 Circuit operation and troubleshooting

It can be seen by the discussion thus far that the picture signal circuits of both Beta and VHS VCRs are very complex. However, the picture circuits are not necessarily complex to troubleshoot and repair. This is especially true in modern VCRs, where most of the picture circuits are contained within a very few ICs. It is relatively easy to trace signals through the circuits (it usually boils down to tracing a few inputs/outputs between ICs). Likewise, 90 percent of picture circuit problems are cured by proper adjustment, followed by replacement of a few ICs, as necessary. Unfortunately, this is not true for the mechanical (tape transport, system control, servo, etc.) functions in VCRs, where most problems arise. That is why we concentrate on the mechanical control functions throughout this book.

1-7.8 Tape loading or threading for the VHS system

Figure 1-29 shows the typical tape loading system for early-model VHS VCRs. Modern VHS units often use a *loading ring* somewhat similar to that described for Beta (and discussed in Chapter 5). Note that the term *tape loading* applies primarily to threading of the tape, and is not to be confused with *cassette loading,* which involves inserting and locking the tape cassette within the VCR from top or front (also described in Chapter 5). However, the term *loading* is used interchangeably in VCR literature.

No matter how it is done, the VHS threading is similar to that shown in Fig. 1-29. As shown, VHS uses the so-called M loading or threading system, since the tape appears to form the letter M when fully threaded. We talk about both top-load and front-load tape transports in Chapter 5.

1-8 THE BASIC VCR SYSTEM CONTROL FUNCTIONS

The system control circuits are not identical for all VCRs. In fact, this is an area where one model of VCR can be quite different from other models. However, most system control circuits have some basic functions in common.

First, the system control circuits coordinate operation of all other VCR circuits during the various operating modes. For example, the system control circuits provide the necessary voltages and signals to keep both the tape and scanner moving during normal record and playback operations, but stop tape movement when the pause operating mode is selected. Since all VCRs do not have the same operating modes, it is not practical to generalize on the functions of the system control circuits.

Another major function of the system control circuits is to provide a fail-safe function that stops operation of the VCR in case of failure, as well as at both ends of the cassette tape. Typical features that stop operation of the VCR include slack tape, excessive moisture on the tape, prolonged operation in the pause mode, and failure of the drum to rotate.

Most system control circuits also provide the control signals for muting audio and video circuits during the fast-forward, fast-rewind, and other operating modes. For example, in Beta VCRs, muting also occurs when there is no CTL signal during playback. This produces a blank picture instead of snow when a blank cassette is played back or at the end of a recording.

Instead of trying to generalize on system control functions here, we refer you to Chapter 14, where we discuss the system control circuits of a modern VCR.

1-9 INSERTING AND REMOVING A VIDEOCASSETTE

As shown in Fig. 1-30, both top-loading and front-loading are used for the cassettes of modern VCRs. Top-loading is still the most popular, but the trend seems to be toward front-loading, to make the VCRs more compact. The following procedures apply generally to all VCRs, early-model and modern.

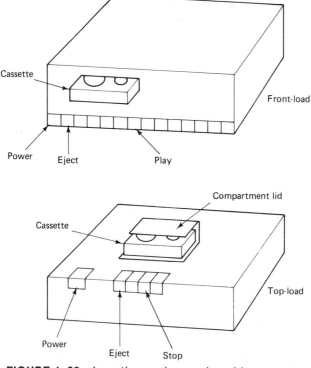

FIGURE 1-30 Inserting and removing videocassettes

However, you should always check the operating literature for any VCR you are servicing. It makes a bad impression on the customer if you cannot find the cassette compartment, especially on the second service call.

To insert a cassette in a typical top-load Beta VCR, you must apply power to the VCR and set the power switch to ON. You then press the EJECT button, which automatically opens and raises the cassette compartment lid. You then install the cassette in the compartment, usually within a holder, and press down on the compartment lid. In about 3 seconds, the VCR is ready to perform any of the normal operations.

When the cassette compartment lid is pressed down, the supply and take-up reels within the cassette engage the drive mechanisms of the *supply* and *take-up drives* within the VCR. These drives are capable of operating in both directions to accommodate both playback/record and rewind functions. The supply and take-up motors are usually capable of operating at various speeds. Typically, the supply and take-up reel motors receive a low voltage during record and playback modes. This low voltage keeps tension on the tape, but the actual tape drive is provided by the capstan and pinch roller. During fast forward and rewind, the supply and takeup motors receive a higher voltage and provide the tape drive. In the case of most top-load VHS machines, the tape is unloaded and returned back inside the cassette when fast forward and rewind are in operation.

For top-load Beta machines, when the lid is closed, the tape is automatically loaded as described in Sec. 1-6.6. For a top-load VHS, tape loading occurs after the lid is closed and the PLAY button is pressed.

To remove a top-load Beta cassette, make sure that the power is turned on and that the VCR is in the STOP mode. On some VCRs, the EJECT button cannot be pressed except in the STOP mode. In other VCRs, the EJECT button can be pressed but does not actuate the circuit unless the VCR is in STOP. Press EJECT, remove the cassette, and close the compartment lid. When a top-load Beta compartment lid is raised by pressing the EJECT button, the tape is automatically loaded as described in Sec. 1-6.6 and the cassette supply and take-up reels disengage from the tape drive motors.

For most top-loading VHS machines, when the STOP button is pressed, the tape is unloaded. The cassette can then be removed by pressing EJECT to release the cassette holder.

To install a cassette in most front-load VCRs, you set the power switch to ON and then insert the cassette partway into the cassette compartment. The cassette is then pulled into place (in and down over the tape load mechanism) by the cassette load motor and gear mechanism. *Note that when a cassette is already installed, a mechanical stop is present. So do not try to force a cassette into the compartment on any front-load VCR!* Once the cassette is in place, the tape is loaded by the loading ring (or whatever tape load system is used) when you press PLAY.

To remove a cassette from most front-load VCRs, you stop the VCR

48 Introduction to Videocassette Recorders

and press EJECT. The tape load mechanism then unloads the tape back into the cassette (if necessary), and the cassette load mechanism operates in reverse to lift the cassette up from the tape load mechanism and partway out of the cassette compartment. Always wait for the cassette to stop moving out from the front panel before trying to pull the cassette free. Also, always remove the cassette carefully.

1-10 RECORD LOCKOUT OR MALERASE FUNCTIONS

Both Beta and VHS cassettes can be reused to record new program material many times. The erase head automatically erases all recorded material on the tape before the tape reaches the rotating video heads (Fig. 1-3). Both Beta

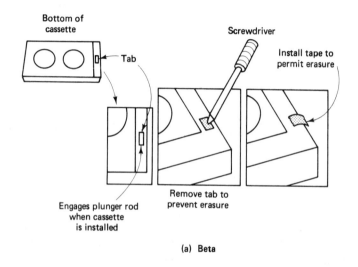

(a) Beta

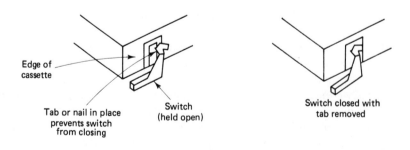

(b) VHS

FIGURE 1-31 Typical record lockout or malerase function

and VHS cassettes have provisions for preventing accidental erasure of recorded material when you want to retain a particular program. These provisions are called *record lockout* or *malerase* or some similar term. In a Beta VCR, the record lockout takes the form of a tab located on the bottom of the cassette, as shown in Fig. 1-31a. In VHS, a similar tab or nail is located on the edge of the cassette, as shown in Fig. 1-31b.

With either system, the tabs engage a plunger rod or switch when the cassette is inserted and the compartment lid is closed. In the case of most Beta systems, the RECORD button cannot be pressed unless the rod is pushed down by the tab. In most VHS systems, the tab prevents a switch from closing (closing of the switch disables record operation). Note that some VHS systems operate in the opposite manner. Either way, if you want to keep a recorded program from being accidentally erased, you break off the tab or nail with a screwdriver. In this way, the record function is disabled and the original recording is preserved.

If you wish to record on a cassette with the tab or nail removed, cover the hole with a piece of tape. The tape actuates the plunger rod or holds the switch, and the record function is normal. During service, it is sometimes necessary to operate the VCR without a cassette installed (for instance, to observe rotation of the tape drive motor). In this case, you can use the tape to hold the plunger rod or switch in place, just as if actuated by the tab or nail.

1-11 FEATURES OF A MODERN VCR

The following are brief descriptions of the features found in a typical modern VCR. The circuit details and mechanical operation of the features are discussed in the appropriate chapters. Keep in mind that the features discussed here are typical and represent a cross section of modern VCRs.

Four-head assembly. Most modern VCRs have four heads (a few models have five heads). There are two basic four-head designs (90° spacing and pair spacing), as shown in Fig. 1-32. We discuss these designs further in Chapter 7. For now, it is sufficient to say that the real advantage of the pair-spacing design is a significant improvement of special effects (still frame, slow-motion, speed search). With conventional 90° spacing, the VCR reads two video images at a time when in still frame or slow motion. If the scene shows motion or camera movement, the separate images are often different. (The jitter normally associated with still-frame images is the result of the picture shifting between the two images.)

Direct-drive (DD) motors. Most modern VCRs replace inefficient belts and pulleys in the tape transport with microcomputer-controlled DD motors. Typically, there is one motor for each of the following functions: drum,

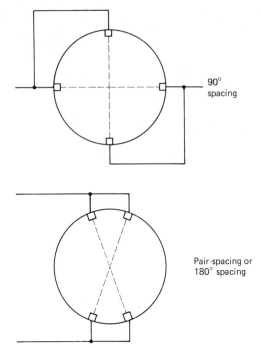

FIGURE 1-32 Basic four-head configurations

capstan, take-up reel, supply reel, tape loading, and (on front-loading models) cassette loading. This direct-drive operation is extremely accurate, quiet, and reliable (and avoids the problems of *belt stretching* so common in early-model VCRs). All of the DD motors are synchronized electronically by the servo and system control functions.

Remote control. Most modern VCRs have some form of remote control. Typically, the remote is of the IR (infrared) wireless type found in modern TV sets. The remote control of the VCR in Fig. 1-1a has 29 functions, including direct access or channel scanning of the 139-channel frequency-synthesis tuning.

139-channel cable-ready, frequency-synthesized tuning. Frequency-synthesized tuning is common for modern VCRs. Microcomputer-controlled electronic tuning locks in the best signal of each channel. In the normal TV mode you have access to VHF channels 2 to 13 and UHF channels 14 to 83. In cable mode you have 12 VHF plus 15 midband, 14 superband, and 28 hyperband, or a total of 69 cable channels. Large front-panel illuminated digits display current VHF, UHF, or cable channels.

Monitor control. The VCR of Fig. 1-1a has one unique feature called *monitor control.* The volume level and mute of the VCR's audio outputs are controlled by the remote, allowing the VCR remote to control the TV audio as well. When the power plug of the TV is plugged into the outlet on the VCR, you can turn the TV set on and off from the VCR remote. This in effect converts the TV into a 139-channel cable-ready set complete with full-function wireless remote. However, not all VCRs have this function, nor is it possible to operate all TVs in this way. (For example, TVs with remote or electronic turn-on can not be turned on and off by the VCR remote.) You should be aware of such characteristics when trying to troubleshoot a feature that does not exist.

Dual-channel audio with Dolby* noise reduction. Most modern VCRs have dual-channel stereo audio, which is ideal for bilingual programming or prerecorded music videotapes (music videos). Also, by using FM tuner outputs of home audio systems, you can combine simulcast FM audio with TV broadcast video on the same videocassette. The dual-channel feature is also particularly effective when adding narration or music to already-recorded video and audio programming. For example, the audio-dub feature of the Fig. 1-1a VCR permits right-channel audio recording while not affecting the left channel or previously recorded video.

The Dolby noise-reduction system provides clean, noise-free audio recordings. Such a system is particularly important when recording in the slower LP and EP modes.

Programmed recording switch. When programmed recording information is entered, you press this switch to set the VCR in the record-ready mode.

Operation mode display. This feature provides an illuminated display of various tape operations.

Counter/timer, counter reset, and memory. These buttons provide control of the time-of-day/digital counter display, allow resetting of the tape counter to zero, and initiate the memory-rewind function.

Remaining time. This feature is used to display accurately the amount of time remaining on the videocassette in any mode. The time indicated is accurate within one minute regardless of tape speed.

---

*Noise reduction system manufactured under license of Dolby Laboratories; Dolby and the double-D symbol are trademarks of Dolby Laboratories Licensing Corporation.

Tape speed. This control sets the recording speed to SP (standard play), LP (long play), and EP (extended play), the three speeds available on most modern VCRs. Note that the playback speeds are automatically detected and switched when the VCR is in play.

Channel selection. These buttons provide automatic or manual up/down scan of VHF/UHF or cable channels.

Tape-function controls. These controls provide smooth, precise front-panel control of all tape functions, including rewind, fast forward, play, stop, record, pause, and speed search (9× in EP, 7× in SP).

Video fine-edit. This feature allows you to make professional-looking transitions between selections with no flutter or perceptible "glitch" at the edit point.

Eight-event, 14-day programmability. A typical modern VCR allows you to program up to two weeks in advance on up to eight different programs. Or you can program the VCR to record the same time and channel each day. The entry of data, start and end times, and channel is simple and straightforward. Also indicated is a 30-minute battery backup that protects the setting in case of accidental power loss.

Tape-tracking controls. Individual tape-tracking controls enable fine adjustment of tape-to-head alignment during normal-speed and slow-motion playback. Note that early-model VCRs often have only one tracking control for all speeds, whereas most modern VCRs have two tracking controls.

Picture control. This feature—rarely, if ever, found on early-model VCRs—provides special video-enhancement circuitry that gives added sharpness and picture quality. The feature allows you to set the desired amount of sharpness and is especially useful for playback of mass-duplicated prerecorded cassettes that may lack detail (copies of copies of copies).

One-touch recording (OTR). This feature automatically sets the VCR into the record mode for 30 minutes at a time. No other settings need be made. Each additional touch of the control adds 30 more minutes (up to two hours) to the recording time. The total time is indicated on the multifunction digital display and counts down to zero as recording progresses.

Video/TV mode. This control is for selection of either the TV reception mode or the VCR mode. However, when playing a videotape, the VCR senses the playback and automatically switches to the video mode. TV/video switching is also provided on the remote, in most cases.

Input selector. This feature allows selection of line, camera, or tuner as the input recording source.

Cable/TV mode and auto/manual tuning. These controls determine whether or not the tuning mode is normal VHF/UHF reception or cable and whether or not tuning control is in an auto-scan or manual mode.

Fine tuning. Normal fine tuning is automatically done by the microcomputer-controlled frequency-synthesizer tuner. However, most modern VCRs also include some form of manual fine-tuning control.

Audio monitor and microphone jacks. Most modern VCRs have dual microphone inputs for live audio recording. Typically, one-channel input is also used in conjunction with the audio-dub function to record live audio in sync with previously recorded video and opposite-channel audio.

1-12 OPERATING CONTROLS, INDICATORS, AND CONNECTORS OF A MODERN VCR

Now that we have reviewed the features of modern VCRs, let us compare them to the controls and indicators of a specific VCR (the Magnavox VR8440BK01). The operating controls and indicators are shown in Fig. 1-33; the connectors are shown in Fig. 1-34. Note that the VCR has a front-panel door which opens to expose operating controls. The VCR can also be operated by the hand-held remote transmitter. Complete function, mode, tuner, time, and tape counter are displayed on a fluorescent display tube. (The use of a single, special-purpose display tube is common for modern VCRs.)

1-12.1 VCR controls and indicators

As shown in Fig. 1-33, the cassette is inserted through the front-panel *cassette compartment door*. You slide the cassette into the VCR until the front-load mechanism draws the cassette in and down automatically. The *power switch* must be on. When the power switch is on, the count 0000 appears on the *indicator panel*. The *wireless remote sensor* receives signals from the wireless remote control.

You push the *EJECT button* to remove the cassette (the cassette is moved partway out of the cassette compartment door by the load mechanism). You push the *DOOR OPEN button* to open the *control panel door*.

The *input signal selector* can be set to *camera* (for camera recording, either stereo or monaural), *line* (for rerecording, audio dub, or video dub), *tuner* (for regular TV recording with monaural sound recording), *audio two-channel* (for simulcast, FM stereo, recording). The *channel up/down keys* are

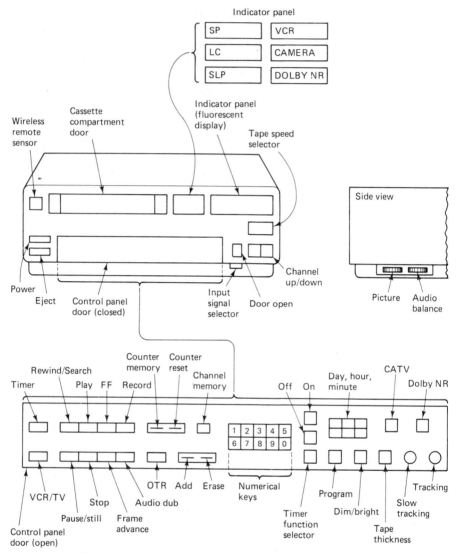

FIGURE 1-33 Typical operating controls and indicators

used to change *preset* channels. The *tape speed selector* determines the length of the program to be recorded (SP/LP/SLP). Note that SLP is often referred to as EP, or *extended play,* on some VCRs (such as the one described in Sec. 1-11).

The six indicators grouped next to the indicator display panel or tube show the following: SP, LP, and SLP tape-speed indicators show the tape speed during both record and playback; the VCR indicator goes on when the control panel VCR/TV selector is set to VCR; the camera indicator goes on

Operating Controls, Indicators, and Connectors of a Modern VCR 55

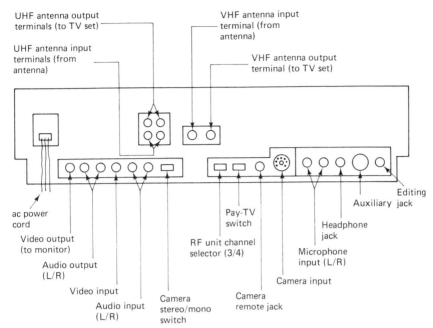

FIGURE 1-34 Typical rear-panel connectors and switches

when the input signal selector is in camera position; the Dolby NR indicator goes on when the Dolby NR switch is on.

The TIMER button is used for unattended recording after programming functions are complete. When on, a *timer indicator* display appears on the indicator panel and the VCR cannot be operated manually.

The REW/SEARCH button is used to rewind tape. When on, a rewind display appears on the indicator panel. Holding the REW/SEARCH button allows you to view the picture in (rapid) reverse.

The PLAY button is used to play back recorded tapes. When on, a play display appears on the indicator panel.

The FF/SEARCH button moves the tape forward rapidly. When on, a fast-forward display appears on the indicator panel. Holding the FF/SEARCH button allows you to view the picture in (rapid) forward.

The REC button is used to start recording. The PLAY button must be pressed at the same time. When both are on, a record function appears on the indicator panel.

The COUNTER MEMORY button is used to stop the tape rewind when the tape counter reaches 0000.

The COUNTER RESET is used to reset the tape counter to 0000. By beginning a recording at 0000, subsequent playback is more convenient.

When the CH MEMORY (channel memory) switch is on, the ADD and ERASE buttons can be used to add or erase TV channels by means of the numerical keys. After presetting the desired channels, the CH MEMORY switch is set to off, and the frequency-synthesized tuning (Chapter 10) takes over.

The ON and OFF buttons are used before setting the start and stop times, respectively, for each program. The DAY, HOUR, and MIN buttons are used to set present time or start/stop times for each program. The corresponding time display appears on the indicator panel.

The CATV mode selector must be set to the desired position (TV, CATV, NOR, HRC, ICC), depending on the cable system in the area. If a cable system is not used, the control is set to TV.

The DOLBY NR switch is set to on for audio noise reduction, as shown by the Dolby NR indicator.

The tracking control is used during playback if the image is partially obscured by bands of noise. When troubleshooting, always set the tracking control to the detent (center) position as a starting point. The tracking control also provides for minor variations between tapes recorded on one machine and played back on another machine. If the physical distance between the control head and video heads is different for the two machines, the playback signals are not synchronized, even though the servo is locked to the CTL signal. This condition can be corrected by physically moving the control/audio head stack in relation to the drum. (That is one of the recommended service adjustment procedures for some VCRs.) However, it is far more practical for the user to operate a front-panel tracking control. The slow tracking control is used for slow-motion playback.

The tape-thickness switch is used to judge remaining tape time from the tape balance display on the indicator panel. When T-160 tape is used, the switch is set to THIN. The switch is set to NORMAL for all other tapes.

The DIM/BRT switch sets the brightness level of the indicator panel.

The PROGRAM button is used to select the desired program number when programming the timer. When the timer function selector is set to CLOCK, the present time is set into the digital clock on the indicator panel. While setting the timer for record, the timer function selector is set to PROGRAM. Except for these two cases, leave the selector in the NORMAL position.

The OTR (one-touch recording) button permits you to make impromptu recordings at any time. Simply select the desired channel and press the OTR button to get 30 minutes to four hours of recording.

When the A DUB (audio dub) and PLAY buttons are pressed simultaneously, sound from another source can be recorded on the same tape in place of the original sound (the original sound is erased automatically).

While viewing a still picture, push the F ADV (frame advance) button to advance the picture one frame at a time. When F ADV is held down, you get slow-motion pictures. When F ADV is on, a corresponding display appears on the indicator panel.

You push the STOP button to stop the tape. You push the PAUSE/STILL button to temporarily stop the tape movement in either the recording or playback mode. During playback, a still picture is produced when PAUSE/STILL is used (you push PAUSE again to release the pause function). When the PAUSE/STILL button is used, corresponding displays appear on the indicator panel.

The VCR/TV selector is set at VCR to monitor video recordings or to view playback, and is set to TV to watch TV or to view one program while recording a different program. When set to VCR, a corresponding display appears on the indicator panel.

The *picture control* (located on the side) is used to make the picture softer or sharper, whichever you prefer. The *audio balance control* selects the desired stereo balance.

The following is a description of the displays appearing on the indicator panel:

A *digital clock* indicates present time (once properly set). A *tape counter* shows the tape count (in units from 0000 to 9999). A *function indicator* shows the operating mode (eject, play, record, rewind, fast foward, pause, still search, stop, frame advance, slow). A *dew indicator* appears if excessive moisture condenses in the VCR (stopping the VCR but not removing power). If this happens, leave the VCR power on and let the VCR stand at room temperature until the dew indicator goes off, indicating that excessive moisture has been removed. Allow about an hour before starting operation. The AFT (automatic fine-tune) display should be on at all times. If not, there may be problems in the tuning circuits (Chapter 10). The *channel display* shows the channel number selected by the numerical keys on the control panel. A *tape-balance indicator* shows the remaining length of tape time by five-minute units. The *timer indicator* appears when the TIMER button is pressed (when on, the VCR cannot be operated manually). A *program number* display shows the program number for timer recording (a W appears for every week recording). An OTR indicator shows when OTR is selected. A *cassette-in* indicator shows when the cassette is in place and ready to record or play back. CATV appears on the indicator panel when the CATV mode selector is set to CATV. A *memory indicator* (M) appears when the COUNTER MEMORY button is set to on. A *dub indicator* (DUB) appears when a dub function is selected.

1-12.2 VCR connectors and rear-panel switches

The connectors and switches, and their related functions, are shown in Fig. 1-34. Note that these connectors and switches are located on the rear panel and are not generally used by the consumer.

2

THE BASICS OF VCR TROUBLESHOOTING AND REPAIR

In this chapter we discuss the basic approaches for troubleshooting and repair of modern VCRs. We cover such areas as safety precautions, test equipment, special tools, cleaning, lubrication, and general maintenance. The chapter concludes with some basic troubleshooting procedures and notes that apply to VCRs of all types, early-model and modern. The troubleshooting procedures for specific sections (servo, video, etc.) of modern VCRs are covered at the end of the related chapter.

Keep in mind that the information in this chapter is general in nature. If you are going to service a particular VCR, get all the service information you can on that VCR. Likewise, if you plan to go into VCR service on a large scale, study all the applicable service literature you can find. Then, when all else fails, you can follow instructions.

2-1 SAFETY PRECAUTIONS IN VCR SERVICE

In addition to a routine operating procedure (for both test equipment and the VCR), certain precautions must be observed during operation of any electronic test equipment. Many of these precautions are the same for all types of test equipment; others are unique to special test instruments, such as meters, oscilloscopes, and signal generators. Some of the precautions are designed to prevent damage to the test equipment or to the circuit where the service operation is being performed. Other precautions are to prevent injury to you.

Where applicable, special safety precautions are included throughout the various chapters of this book.

2-1.1 General safety precautions

The following general safety precautions should be studied thoroughly and then compared to any specific precautions called for in the test equipment service literature and in the related chapters of this book.

Warning symbols on test equipment. There are two standard international operator warning symbols found on some test equipment. One symbol, *a triangle with an exclamation point at the center,* advises the operator to refer to the operating manual before using a particular terminal or control. The other symbol, *a zigzag line simulating a lightning bolt,* warns the operator that there may be dangerously high voltage at a particular location, or that there is a voltage limitation to be considered when using a terminal or control. Always observe these warning symbols. Unfortunately, the use of symbols is not universal, particularly on older test equipment.

Metal cases. Many service instruments are housed in metal cases. These cases are connected to the ground of the internal circuit. For proper operation, the grounded terminal of the instrument should always be connected to the ground of the VCR being serviced. *Make certain* that the VCR chassis *is not* connected to either side of the ac line or to any point above ground, using the leakage current check of Sec. 2-1.2.

High voltages. Remember that there is always danger in servicing VCRs that operate at hazardous voltages, especially if you pull off covers with the power cord connected. Fortunately, most VCR circuits operate at potentials well below the line voltage, since the circuits are essentially solid state. However, a line voltage of 120 V is sufficient to cause serious shock and possibly death! Always make some effort (such as reading the service literature) to familiarize yourself with the VCR before service, bearing in mind that line voltages may appear at unexpected points in a defective VCR.

Remove power. It is good practice to remove power before connecting test leads to high-voltage points. It is preferable to make all service connections with the power removed. Since this is generally impractical, be especially careful to avoid accidental contact with player circuits. Keep in mind that even low-voltage circuits may be a problem. For example, a screwdriver dropped across a 12-V line in a solid-state circuit can cause enough current to burn out a major portion of the VCR, possibly beyond repair. Of course, that problem is nothing compared to the possibility of injury to yourself! Working with one hand away from the VCR, and standing on a properly insulated floor, lessen the danger of electrical shock.

Capacitors may store a charge large enough to be hazardous, although generally not in solid-state circuits. Discharge capacitors before attaching test leads. (Please make sure that you have turned off the power before you discharge the capacitors!)

Remember that leads with broken insulation offer the additional hazard of high voltages appearing at exposed points along the leads. Check test leads for frayed or broken insulation before working with them. To lessen the danger of accidental shock, disconnect test leads immediately after the test is complete.

Remember that the risk of severe shock is only one of the possible hazards. Even a minor shock, or touching a hot spot, can put you in danger of more serious risks, such as a bad fall or contact with a source of higher voltage.

The experienced service technician guards continually against injury and does not work on hazardous circuits unless another person is available to assist in case of accident. Even if you have considerable experience with test equipment used in service, always study the service literature of any instrument with which you are not thoroughly familiar. Use only shielded leads and probes. Never allow your fingers to slip down to the metal probe tip when the probe is in contact with a hot or live circuit.

Avoid vibration and mechanical shock. Most electronic test equipment is delicate. The mechanical portions of a VCR are vulnerable to any kind of shock or vibration. Not only can the mechanical parts be damaged, but they can also be thrown out of adjustment by rough handling. This is especially true of the scanner and video heads. Although the heads are designed to be rotated continuously at 1800 rpm and to be in contact with magnetic tape, they are not designed to be in contact with the tips of screwdrivers, Allen wrenches, and the like.

Study the circuit being serviced before making any test. Try to match the capabilities of the test instrument to the circuit being serviced. For example, if the circuit under test has a range of measurements to be made (ac, dc, RF, modulated signals, pulses, or complex waves) it is usually necessary to use more than one instrument. Most meters measure dc and low-frequency signals. If an unmodulated RF carrier is to be measured, use an RF probe. If the carrier to be measured is modulated with low-frequency signals, a demodulator probe must be used. If pulses, square waves, or complex waves are to be measured, a peak-to-peak meter can possibly provide meaningful indications, but an oscilloscope is the logical instrument. If the problem is one of monitoring the digital logic pulses associated with the microcomputer/

microprocessor found in the system control of most modern VCRs, you must use digital test equipment such as logic probes and pulsers. Or you can try a really novel approach and use the test instrument recommended in the VCR service literature!

2-1.2 Leakage current tests

Before placing a VCR in use (for service or normal home use), it is recommended that you measure possible leakage current. Such leakage indicates that the metal parts of the VCR are in electrical contact with one side of the power line. If the leakage problem is severe, it can result in damage to the player or possible shock to anyone touching the exposed metal parts. There are two recommended leakage current tests: *cold check* and *hot check*.

Cold check. With the ac plug removed from the 120-V source, place a jumper across the two ac plug prongs. Turn the ac power switch on. Using an ohmmeter, connect one lead to the jumpered ac plug and touch the other lead to each exposed metal part (metal cabinet, screw heads, metal overlays, control shafts, etc.), particularly any exposed metal part having a return path to the chassis. Such parts should have a minimum resistance reading of 1 MΩ. Any resistance below this value indicates an abnormality that requires corrective action. Exposed metal parts not having any return path to the chassis (an "infinity" reading) shows an open circuit. Generally, any reading above about 5 to 6 M should be suspect. Possible exceptions are the F-type connectors that connect the VCR to the antenna, cable input, or TV set. The center conductors of such connectors often read "open circuit" when the VCR is in various operating modes.

Hot check. Using the diagram of Fig. 2-1 as a reference, measure ac leakage current with a milliammeter. Leave switch S1 open and connect the VCR power plug to the test connector. Immediately after connecting the VCR, measure any leakage current with switch S2 in both positions. Set the player switches (at least the power switch) to ON when making the leakage current measurements. Now close switch S1 and immediately repeat the leakage current measurements in both positions of switch S2 (and with the VCR switches on). Allow the VCR to reach normal operating temperature and repeat the leakage current tests. In any of these tests the leakage current should not exceed about 0.5 mA (for a typical VCR).

If possible, check both the TV and antenna to be used with the VCR for possible leakage currents. Use the same procedure as described for the VCR (Fig. 2-1). To avoid shock hazards, do not connect a VCR to any TV, antenna, cable, or accessory that shows excessive leakage current.

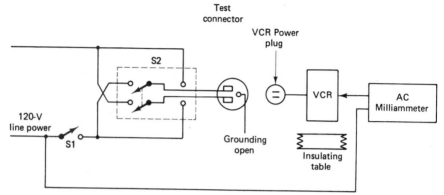

FIGURE 2-1 "Hot check" circuit for leakage current tests

2-1.3 Basic handling and service precautions

The following precautions apply to all types of VCRs and are to be observed *in addition* to any precautions described in the service literature.

Handling and storage. Avoid using the VCR in the following places: extremely hot, cold, or humid places; dusty places; near appliances generating strong magnetic fields; placed subject to vibration; poorly ventilated places. Do not block the ventilation openings. Do not place anything heavy on the VCR. Do not place anything that might spill on the VCR top cover. Use an accessory cover (if available) to prevent dust and dirt from accumulating on the VCR. Use the VCR in the horizontal (flat) position only. Do not lubricate VCR motors or any point not recommended for lubrication. Remove any excess lubricant.

When reassembling any VCR, always be certain that all protective devices (nonmetallic control knobs, shield plates, etc.) are put back in place. When service or testing is required on any VCR, observe the original lead dress (wire routing, etc.). Pay particular attention to parts that show any evidence of overheating or other electrical or mechanical damage. If you must transport a VCR, avoid violent shocks to the machine during both packing and transportation. Before packing, be sure to remove any cassettes from the VCR. Always look for any packing or shipping devices supplied with the VCR (such as shipping screws that must be tightened, etc.).

Moisture condensation. Be very careful to avoid the effects of moisture on a VCR. Moisture condensation on the scanner and video heads, probably the most critical parts of any VCR, cause damage to the tape. Avoid using a VCR immediately after moving it from a cold place to a warm place, or soon after heating a room where it was very cold. Otherwise, the water vapor in warm air will condense on the still-cold video head scanner and tape guides and may cause damage to the tape or VCR or both.

Because of the moisture-condensation problem, most VCRs are provided with a moisture-condensation protection circuit. A moisture sensor (usually called the *dew sensor*) is mounted near the video heads within the VCR. If moisture exceeds a safe level, the sensor triggers a circuit in the system control and removes power to the tape drive and scanner motors. Operation of dew-sensor circuits is discussed in Chapter 14.

Interlocks. Do not defeat any type of interlock on a VCR. If you must override an interlock during service (try to avoid this), *do not permit the VCR to be operated by others without all protective devices correctly installed and functioning. Servicers who defeat safety features or fail to perform safety checks may be liable for any resulting damage.*

Design alterations. Do not alter or add to the mechanical or electrical design of a VCR. Design alterations, including (but not limited to) addition of auxiliary audio and/or video output connections, cables, accessories, and so on, might alter the safety characteristics of the VCR and create a hazard to the user. Any design alterations or additions may void the manufacturer's warranty and may make the servicer responsible for personal injury or property damage resulting therefrom.

Product safety notices. Many electrical and mechanical parts in VCRs have special safety-related characteristics, some of which are often not evident from visual inspection, nor can the protection they give necessarily be obtained by replacing the parts with components rated for higher voltage, wattage, and so on. The manufacturers often identify such parts in their service literature. One common means of such identification is by *shading on the schematic and/ or parts lists,* although all manufacturers do not use shading or limit such identification to shading. For example, RCA goes one step further than most manufacturers and uses a *dark black pattern* on those areas of their printed circuit board copper patterns that require special care in repair. Always be on the alert for any special product safety notices, special parts identification, and so on. Use of a substitute part that does not have the *same safety characteristics* (not just the same electrical or mechanical characteristics) might create shock, fire, and/or other hazards. A simple way to solve the problem is to use the part recommended in the service literature!

Good electronic service practices. The author assumes that you are already familiar with good electronic service practices (removing the power cord before replacing circuit boards and modules, installing heat sinks as required on solid-state devices, connecting test-instrument ground leads to chassis ground before connecting the test instrument, and so on). The author also assumes that you can handle electrostatically sensitive (ES) devices (such as FETs, MOS chips, etc.); that you can solder and unsolder ICs, transistors, diodes, and so on; and that you can repair circuit-board copper foil as needed. If any of these seem unfamiliar to you, please please do not attempt to service any VCR, especially the author's!

Copyright problems. Many of the programs broadcast by television stations and cable networks are protected by copyright, and federal law imposes strict penalties for copyright infringement. Some motion picture companies have taken the position that home recording for noncommercial purposes is an infringement of their copyrights. The U.S. Supreme Court has ruled otherwise. However, there are laws before the U.S. Congress to protect unauthorized recording of TV broadcast material. Likewise, the recording—or even the viewing—of cable broadcasts without authorization (without paying for the cable service) is against the law. So until all these laws have been sorted out, a VCR used to record copyrighted material should be operated at the user's own risk.

2-2 TEST EQUIPMENT FOR VCR SERVICE

The test equipment used in VCR service is basically the same as that used in TV service and in other fields of electronics (particularly record players, stereos, etc.). That is, most service procedures are performed using meters, signal generators, oscilloscopes, frequency counters, power supplies, and assorted clips, patch cords, and so on. Theoretically, all VCR service procedures can be performed using conventional test equipment, provided that the oscilloscopes have the necessary gain and bandpass characteristics, that the signal generators cover the appropriate frequencies, and so on.

For these reasons, we do not go into full details on VCR test equipment in this book. If you have a good set of test equipment suitable for conventional TV and audio work, you can probably service any VCR. Possible exceptions are digital test instruments (probes and pulser) required for troubleshooting the system-control sections of modern VCRs. Instead of full details, we describe the essential characteristics that test equipment must have for good VCR service.

2-2.1 Signal generators

As a minimum, you should have a conventional TV *sweep/marker generator* and an *analyst/pattern generator* for noncolor tests and adjustments. An *NTSC-type color generator* is essential for VCR service. The *keyed-rainbow color generator* provides some of the signals you need, but not all. The two signals most often required for VCR service are the *standard NTSC bar pat-*

Test Equipment for VCR Service   65

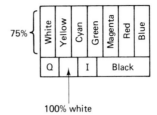

FIGURE 2-2 Standard NTSC bar pattern (75%) with a −IWQ signal occupying the lower quarter of the pattern

tern with a -IWQ signal occupying the lower quarter of the pattern, as shown in Fig. 2-2, and the *five-step linear staircase pattern,* similar to that shown in Fig. 2-3, with selectable chroma levels.

Most NTSC-type color generators provide these required signals (and much more). For example, the B&K Precision, Dynascan Corporation Model 1251 NTSC Color-Bar Generator, shown in Fig. 2-4, is typical of present-day NTSC generators. In addition to supplying the -IWQ and linear staircase patterns, the model 1250 provides many other patterns including dot, crosshatch, dot hatch, center cross, and full-color raster. The raster colors can be displayed by using three color raster buttons individually or in combination (to produce red, blue, green, yellow, cyan, magenta, white, and black rasters). The instrument also generates an RF output on either channel 3 or 4 or at the

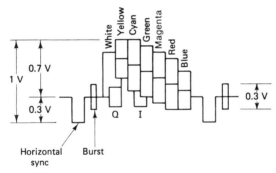

(a) Color bar signal waveform

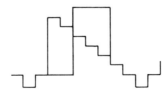

(b) Corresponding five-step linear staircase pattern

FIGURE 2-3 Standard NTSC color bar (75%) signal waveform with corresponding five-step linear staircase pattern

FIGURE 2-4 Model 1251 NTSC color bar generator (Photo Courtesy of DYNASCAN CORPORATION/B&K PRECISION)

standard TV IF frequency of 45.75 MHz. These carriers can be modulated by any of the patterns. In addition to the video patterns, the instrument generates a 4.5-MHz sound carrier with 1- or 3-kHz modulation, or with external modulation, a vertical or horizontal trigger output pulse, and a chroma subcarrier signal of 3.579545 MHz.

2-2.2 Oscilloscopes

Ideally, an oscilloscope for VCR service should be *dual-trace* and must have a *triggered sweep*. The bandwidth must be 5 to 6 MHz, with 10 MHz preferred. The minimum sweep time should be 0.1 μs per division, although a 0.2-μs sweep will probably do the job. TVV and TVH sweeps are helpful but not essential.

2-2.3 Meters

The meters used for VCR service are essentially the same as for all other electronic service fields.

2-2.4 Frequency counters

The two most common uses for a frequency counter in VCR service are (1) to check or adjust the various 3.58-MHz oscillators in the chroma record and playback circuits, and (2) measurement of the servo system timing. Most frequency counters have sufficient frequency range to measure the 3.58-MHz signals. There are low-cost portables that measure up to 30 MHz at an accuracy within 10 Hz. It is the low end of the frequency range that can be a problem.

Many of the servo system signals are in the 30-Hz range (the video head scanner speed, for example). Inexpensive counters do not go down to that frequency. Even if they do cover low frequencies, they are not sufficiently accurate. The accuracy of a frequency counter is set by the *stability of the time base* rather than the readout. The readout is typically accurate to within ±1 count. The time base of a typical shop counter is 10 MHz and is stable to within 10 ppm (parts per million) or 100 Hz. The time base of a precision lab counter can be in the order of 4 MHz and is stable to within 1 ppm or 4 Hz. (Accuracy is not to be confused with resolution. The resolution of an electronic counter is set by the number of digits in the readout.)

Obviously, an accuracy of 100 Hz is not sufficient to measure 30-Hz signals. One way to overcome this problem is to use the *period function,* as is often done when measuring turntable speeds. (Unfortunately, many inexpensive counters do not have a period function.) Period is the inverse of frequency (period = 1/frequency). When period is measured on a counter, the unknown input signal controls the counter timing gate, and the time-base frequency is counted and read out. For example, if the time-base frequency is 1 MHz, the indicated count is in microseconds. (A count of 333 indicates that the gate has been held open for 333 µs.) In effect, the time-base accuracy is divided by the time period. Thus, for 30-Hz signals, where the time period is approximately 1/30 s, the 100-Hz accuracy is increased to 3.3 Hz (100/30), and the 4-Hz accuracy is increased to 1.3 Hz 4/3. Of course, the period count must be divided into 1 to find the frequency.

The accuracy of frequency counters used for VCR service should be checked periodically, at least every six months. Always follow the procedures recommended in the counter service instructions. Generally, you can send the instrument to a calibration shop or to the factory, or you can maintain your own frequency standard. (The latter is generally not practical for most VCR service shops.)

No matter what standard is used, keep in mind that the standard must be more accurate, and have better resolution, than the frequency-measuring device, just as the counter must (theoretically) be more accurate than the VCR.

There are two methods for checking frequency counters. One way is to check the counter against the frequency information broadcast by U.S. Government station WWV. The procedure requires a communications-type receiver, preferably with an S-meter; it is described in my best-selling *Handbook of Practical CB Service* (Englewood Cliffs, N.J.: Prentice-Hall, Inc., 1978).

A simpler method for checking the counter is to monitor the 3.58-MHz oscillator in a color TV. This oscillator is locked in frequency to a color broadcast at a frequency of 3.579545 MHz. The TV oscillator remains locked at this frequency, even though the phase and color hue may shift. So the counter should read 3.579545 MHz when the oscillator frequency is measured. Of course, a *seven-digit counter* is required to get the full frequency resolution.

2-2.5 Oscilloscope and meter probes

Practically all meters and oscilloscopes used in VCR service operate with some type of probe. In addition to providing for electrical contact with the VCR circuit under test, probes serve to modify the voltage being measured to a condition suitable for display on an oscilloscope or readout on a meter. Typical probes for VCR service include *basic, low-capacitance, RF, and demodulator*.

2-2.6 Digital logic test equipment

Virtually all modern VCRs use some form of microcomputer and/or microprocessor in the system control. Often, several microprocessors are used. While it is possible to monitor most microprocessor signals with an oscilloscope, there are special test instruments for digital troubleshooting. As a minimum, you should have (and know how to use) a *logic probe*. A *logic pulser* and a *current tracer* are also useful, but not essential. (There are also *logic comparators, logic analyzers,* and *signature analyzers,* but none of these are absolutely essential for VCR service.

We do not go into the use of digital test equipment in this book. To do so requires too much space. If you are not familiar with digital troubleshooting techniques, your attention is directed to my best-selling *Handbook of Advanced Troubleshooting* (Englewood Cliffs, N.J.: Prentice-Hall, Inc., 1983).

2-2.7 Industrial video receiver/monitor and monitor-type TV sets

If you are planning to go into VCR service on a full-scale basis, you should consider a *receiver/monitor* such as that used in *studio or industrial video work*. These receiver/monitors are essentially TV receivers, but with video and audio inputs and outputs brought out to some accessible point (usually on the front panel). There are also *monitor-type TV sets* designed specifically for VCRs, videodisc players, and video games.

The output connections make it possible to monitor broadcast video and audio signals as they appear at the output of a TV IF section (the so-called *baseband signals,* generally in the range of 0 to 4.5 MHz, at 1 V peak-to-peak for video and 0 dB, or 0.775 V, for audio). These output signals from the receiver/monitor can be injected into the VCR at some point in the signal flow past the tuner IF. (Note that monitor-type TV sets do not generally provide baseband outputs.)

The input connections on either a receiver/monitor or a monitor-type TV make it possible to inject video and audio signals from the VCR (before they are applied to the RF output unit) and monitor the display. Thus, the baseband output of the VCR can be checked independently from the RF unit.

If you do not want to go to the expense of buying an industrial receiver/monitor or a monitor-type TV, you can use a standard TV to monitor the VCR. Of course, with a TV receiver, the VCR video signals are used to modulate the VCR RF unit. The output of the RF unit is then fed to the receiver antenna input. Under these conditions it is difficult to tell if faults are present in the VCR video or in the VCR RF unit. Similarly, if you use an NTSC generator for a video source, the generator output is at an RF or IF frequency, not at the baseband video frequencies.

If you use a TV receiver as a monitor, adjust the vertical height control to *underscan the picture*. This makes it easier to see the video switching point in relation to the start of vertical blanking.

2-3 TOOLS AND FIXTURES FOR VCR SERVICE

Figure 2-5 shows some typical tools and fixtures recommended for field service of VCRs. These tools are available from the VCR manufacturer. In some cases, complete tool kits are made available. There are other tools and fixtures used by the manufacturer for both assembly and service of VCRs. These factory tools are not available for field service (not even to factory service centers, in some cases). This is the manufacturer's subtle way of telling service technicians that they should not attempt any adjustments, electrical or mechanical, not recommended in the service literature.

The author strongly recommends that you take this subtle hint! He has heard many horror stories from factory service people concerning the "disaster area" VCRs brought in from the field. Most of these problems are the result of tinkering with mechanical adjustments—although there are some technicians who can destroy a VCR with a simple electrical adjustment. One effective way to avoid this problem is to use only recommended factory tools and perform only recommended adjustment procedures. Always remember the old electronics rule, "When all else fails, follow instructions."

2-3.1 Alignment tapes

Use of the tools shown in Fig. 2-5 is described through the remaining chapters. One tool merits some discussion here. Most VCR manufacturers provide an alignment tape as part of their recommended tools. An alignment tape is housed within a standard cassette and has several very useful signals recorded at the factory using very precise test equipment and signal sources. Although there is no standardization, a typical alignment tape contains audio signals at low and high frequencies, such as 333 Hz and 7 kHz, an RF sweep, a black-and-white (monoscope) signal or pattern, and NRSC color bar signals. If you intend to service one type of VCR extensively, you would do well to invest in the recommended alignment tape.

70   The Basics of VCR Troubleshooting and Repair

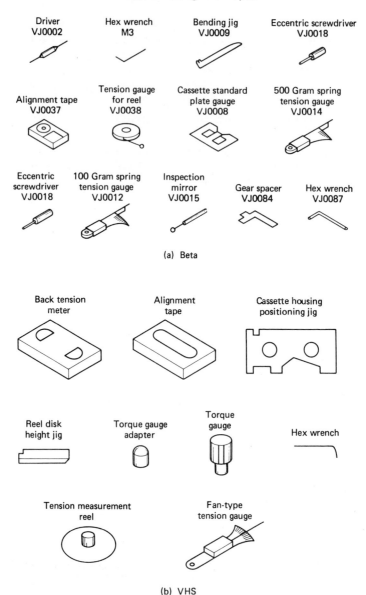

FIGURE 2-5 Typical tools, jigs, and fixtures for Beta and VHS mechanical sections

A typical use for the audio signals recorded on the alignment tape is to check overall operation of servo speed and phase control systems. For example, if the frequency of an audio playback is exactly the same as recorded (or within a given tolerance), and remains so for the entire audio portion of the tape, as checked on a frequency counter, the servo control systems, both speed and phase, must be functioning normally. If there are any mechanical variations or variations in servo control that produce wow, flutter, jitter, and so on, the audio playback varies from the recorded frequency.

If you do not want to invest in a factory alignment tape, or if you do not want to wear out an expensive factory tape for routine adjustments (alignment tape deteriorates with continued use), you can make up your own alignment tape or "work" tape using a blank cassette. The TV stations in most areas broadcast color bars before or after regular programming. (Use the VCR timer for convenience.) These color bars can be recorded using a VCR known to be in *good operating condition*. Any stationary color pattern with vertical lines (such as the 100% white color bar that extends down to the bottom of the screen, as shown in Fig. 2-2) is especially useful. If you have access to a factory tape, you can duplicate it on your own work tape. Of course, make certain to use a known good VCR when making the duplication.

2-3.2 Miscellaneous tools

In addition to the special tools described thus far, the mechanical sections of most VCRs can be disassembled, adjusted, and reassembled with common hand tools such as wrenches and screwdrivers. Keep in mind that most VCRs are manufactured to Japanese *metric standards,* and your tools must match. For example, you will need metric-sized Allen wrenches and Phillips screwdrivers with the Japanese metric points.

Since VCRs require periodic cleaning and lubrication, you will also need tools and applicators to apply the solvents and lubricants (cleaner sticks for the video heads, etc.). Always use the recommended cleaners, lubricants, and applicators, as discussed in Sec. 2-4.

While on the subject of cleaning, you should be aware of a special *cleaning cassette,* also known as a *lapping cassette.* Such cassettes contain a nonmagnetic tape coated with an abrasive. The idea is to load the lapping cassette and run the abrasive tape through the normal tape path (across the video heads, around tape guides, etc.) *for a few seconds.* This cleans the entire tape path (especially the video heads) quite thoroughly. However, prolonged use of a lapping tape can result in damage (especially to the video heads). The author has no recommendation regarding cleaning tape. If lapping cassettes are used, always follow the manufacturer's recommendations, and never use any cleaning tape for more than a few seconds.

2-4 CLEANING, LUBRICATION, AND GENERAL MAINTENANCE

Figure 2-6 shows the recommended maintenance timetable for a typical VCR; it applies generally to both early-model and modern VCRs. However, *never lubricate or clean any part not recommended by the manufacturer.* Most VCRs

The Basics of VCR Troubleshooting and Repair

Names of components \ Intervals (hours)	500	1000	1500	2000	2500	3000	3500	4000	5000
Video heads	△	△	△	△	△	△	△	△	△
Audio/control head	△	△	△	△	△	△	△	△	△
Pinch roller	△	△	△	△	△	△	△	△	△
Erase head	△	△	△	△	△	△	△	△	△
Supply reel				△▲				△▲	
Take-up reel				△▲				△▲	
F Fwd roller		▲		△▲		△▲		△▲	△▲
Clutch pulley				▲		▲		▲	▲
REW idler		▲		△▲		△▲		△▲	△▲
F FWD roller		△		△		△		△	△
Capstan assembly		△		△		△		△	△
Loading gear		△▲		△▲		△▲		△▲	△▲

△ Cleaning

▲ Lubrication

FIGURE 2-6 Recommended maintenance timetable

use sealed bearings that do not require lubrication. A drop or two of oil in the wrong places can cause damage. Clean off any excess or spilled oil. In the absence of a specific recommendation, use a light machine oil such as sewing machine oil.

Although there are spray cans of head cleaner, most manufacturers recommend alcohol and cleaning sticks or wands for all cleaning. Methyl alcohol does the best cleaning job but can be a health hazard. Isopropyl alcohol is usually satisfactory for most cleaning.

We do not go into cleaning and lubrication in any detail here. This is one area where the instructions found in VCR service manuals are usually good. Follow these instructions! In the absence of any instructions, and as a general procedure for all VCRs brought in for service, use the following to clean the heads and tape path.

2-4.1 Video head cleaning

1. Turn the power switch to OFF or pull out the power cord.
2. Rotate the video head drum by hand to a position convenient for cleaning the video head, as shown in Fig. 2-7. Moisten a cleaner stick with alcohol, lightly press the buckskin portion of the stick against

Cleaning, Lubrication, and General Maintenance   73

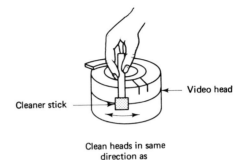

FIGURE 2-7 Video-head cleaning

the video head drum, and move the head by turning the drum back and forth.

3. Clean all four or five heads in the same way.

Caution: Do not move the cleaner stick vertically while in contact with the heads. Always clean the heads in the same direction as the tape path. Cleaning across the tape path can damage the heads.

2-4.2 Audio control, full-erase, and audio-erase head cleaning

Moisten the cleaner stick with alcohol, press the stick against each head surface, and clean the heads by moving the stick horizontally, as shown in Fig. 2-8.

2-4.3 Tape path cleaning

Clean the drum surface and each surface of tape guides with a soft cloth moistened with alcohol. When cleaning the drum surface, be careful not to touch

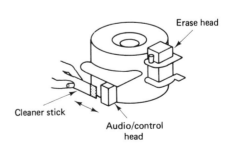

FIGURE 2-8 Audio/control and erase head cleaning

the video head with the cleaning cloth. Rotate the video head drum by hand to move the head away from the spot to be cleaned.

2-5 THE BASIC TROUBLESHOOTING APPROACH

It is assumed that you are already familiar with the basics of electronic troubleshooting, including solid-state troubleshooting. If not, and you plan to service VCRs, you are in terrible trouble. Your attention is directed to my best-selling *Handbook of Advanced Troubleshooting*.

In the case of a VCR, there are seven basic functions required for troubleshooting and repair.

First, you must study the VCR using service literature, user instructions, schematic diagrams, and so on, to find out how each circuit works when operating normally. In this way you know in detail how a given VCR should work. If you do not take the time to learn what is normal, you cannot tell what is abnormal. For example, some VCRs simply produce a better recording than other VCRs, even in the presence of poor TV broadcast signals. You can waste hours of precious time (money) trying to make the inferior VCR perform like a quality set if you do not know what is normal operation.

Second, you must know the function of, and how to manipulate, all VCR controls and adjustments. It is also assumed that you know how to operate the controls of the TV set used to monitor the VCR playback. An improperly adjusted monitor TV can make a perfectly good VCR appear bad. It is difficult, if not impossible, to check out a VCR or TV without knowing how to set the controls. Also, as VCRs and TVs age, readjustment and realignment of critical circuits are often required.

Third, you must know how to interpret service literature and how to use test equipment. Along with good test equipment that you know how to use, well-written service literature is your best friend. In general, VCR service literature is excellent as far as procedures (operation, adjustment, disassembly, reassembly) and illustrations (drawings or photographs) are concerned. Unfortunately, VCR literature is often weak when it comes to descriptions of how circuits operate, why the circuits are needed, and so on (the theory of operation). The "how it works" portion of much VCR literature is somewhat skimpy or simply omitted. I have been told that the reason for such omissions is that you and everyone knows the Beta and VHS system theory, and that most circuits are contained in ICs (and therefore cannot be reached). That is why we concentrate so heavily on the "how it works" aspect of VCRs in the remaining chapters of this book.

Fourth, you must be able to apply a systematic, logical procedure to locate troubles. Of course, a logical procedure for one type of VCR is quite illogical for another. For example, it is quite illogical to check operation of freeze frame or extended play on a VCR not so equipped. However, it is quite logical to check the video, audio, and control tracks on all VCR tapes. We discuss logical troubleshooting procedures for the various sections of a VCR at the end of the corresponding chapter.

Fifth, you must be able to analyze logically the information of an improperly operating VCR. The information to be analyzed may be in the form of performance, such as the appearance of the picture on a known good monitor TV, or may be indications taken from test equipment, such as waveforms monitored with an oscilloscope. Either way, it is your analysis of the information that makes for logical, efficient troubleshooting.

One problem in analyzing a VCR during service is the fact that VCRs must be played back through a TV. If the TV is defective or improperly adjusted, the VCR may appear to have troubles. One practical way to confirm troubles in the VCR is to record a few minutes of a broadcast while watching the program carefully. Then play back the recorded material and switch between the broadcast and playback, comparing the playback quality to that of the broadcast.

Another practical suggestion for evaluation of trouble symptoms *in the shop* is to have at least one TV of known quality. All VCRs passing through the shop can be compared against the same standard. Of course, the ultimate VCR monitor is the industrial receiver/monitor described in Sec. 2-2.7. A next-best alternate is a monitor-type TV, also discussed in Sec. 2-2.7.

The obvious first move in analyzing a VCR, once you have determined that there is a problem with the VCR and not with the TV or antenna, is to play back an alignment tape or a program tape of known quality. This effectively splits the VCR circuits in half. (If you are a regular reader of my books, you know that the half-split is a standard technique in electronic troubleshooting. If you are not a regular reader, you should be!)

If the playback is satisfactory (with a known good tape), it is reasonable to assume that the heads, playback circuits, and RF unit (or modulator) are good. It is also reasonable to assume that the servo, system control, mechanical section, and power supply are good. If the playback is not satisfactory, all of the foregoing circuits are suspect.

The next logical move depends on the results of the first tests. If the playback is good, the tuner/IF and record circuits are suspect. If the VCR has *video and output input connectors,* these can be used to localize the trouble to either the tuner/IF or record circuits. You can use *signal injection* and apply video/audio signals (from an NTSC-type generator) to the video/audio connectors, record the generator signals, and then play back the recorded material.

If the VCR is capable of playing back signals applied at the video/audio input connectors, but not when the signals are applied at the antenna, the problem is in the tuner/IF. If playback is not satisfactory with signals at the video/audio connectors, but a good tape can be played back, the fault is in the record circuits.

If the playback is not good with an alignment tape or other known good tape, the playback circuits and the RF unit or modulator are suspect. If the VCR has *video/audio output connectors,* these can be used to localize the trouble to either the RF unit or playback circuits. You can use *signal tracing* and monitor the audio and video signals (from the good tape) at the connectors within an oscilloscope (for video) and frequency meter or multimeter (for audio).

If the VCR is capable of playing back known good recorded material and producing the correct signals at the video/audio output connectors, the problem is in the RF units. If the playback is not satisfactory at the video/audio output connectors, with a known good tape, the trouble is in the playback circuits.

Sixth, you must be able to perform complete checkout procedures on a VCR that has been repaired. Such checkout may be only a simple operation, such as selecting each operating mode in turn and switching through all channels. At the other extreme, the checkout can involve complete readjustment of the VCR, both electrical and mechanical. In any event, some checkout is required after any troubleshooting.

One practical reason for the checkout is that there may be more than one trouble. For example, an aging part may cause high current to flow through a resistor, resulting in burnout of the resistor. Logical troubleshooting may lead you quickly to the burned-out resistor. Replacement of the resistor can restore operation. However, only a thorough checkout can reveal the original high-current condition that caused the burnout.

Another reason for after-service checkout is that the repair may have produced a condition that requires readjustment. A classic example of this is where replacement of the video heads often requires readjustment of both electrical and mechanical components on VCRs.

Seventh, you must be able to use proper tools to repair the trouble. As discussed at the beginning of this chapter, VCR service requires all of the common hand tools and test equipment found in TV service, plus many special tools, jigs, and fixtures that are unique to the particular VCR. As a minimum, you must have (and be able to use) *tension gauges* and various *metric tools.* These items are often not familiar to the average TV service technician (unless the technician also happens to service tape recorders, stereo decks, etc.).

In summary, before troubleshooting any VCR, ask yourself these questions: Have I studied all available service literature to find out how the VCR works (including any special features such as remote control, search, freeze frame, etc.)? Can I operate the VCR control properly? Do I really understand the service literature and can I use all required test equipment (especially digital test equipment) or tools properly? Using the service literature and/or previous experience on similar VCRs, can I plan out a logical troubleshooting procedure? Can I analyze logically the results of operating checks as well as checkout procedures involving test equipment? Using the service literature and/or experience, can I perform complete checkout procedures on the VCR, including realignment, adjustment, and so on, if necessary? Once I have found the trouble, can I use common hand tools to make the repairs? If the answer is no to any of these questions, you simply are not ready to start troubleshooting any VCR, modern or otherwise. You had better start studying!

2-6 BASIC VCR TROUBLESHOOTING PROCEDURES

Before we get into the detailed troubleshooting/repair notes of Sec. 2-7, or the section-by-section troubleshooting found in the remaining chapters, let us review some simple, obvious steps to be performed before you tear into the VCR with soldering tool and hacksaw. These steps involve such things as checking for proper connections, adjusting the TV, operating the control in proper sequence, and so on.

2-6.1 Simple diagnostic procedure

If the video playback (TV picture) is bad, set the program select switch on the VCR to TV. Check picture quality for each TV channel (using the TV channel selector). If the picture quality is still bad, check for defective antenna connections. For example, the antenna may be defective, the VHF and UHF connections may be reversed, the F-type connector plug may be improperly connected, the center wire in the coaxial cable may be broken, or the TV 75Ω/300Ω switch may be in the wrong position. Also check the TV fine tuning.

If the TV picture is good when the program select switch is set to TV, but video playback is not good, set the program select switch to VCR, turn the TV to the inactive channel (3 or 4), and check reception on each channel by turning the VCR channel selector. If picture quality is bad or there is no picture on all channels, it is possible that the TV fine tuning is not properly adjusted. If the problem appears only on certain channels, the VCR fine tuning is suspect, as is the VCR tuner.

If picture quality is good when viewing a TV broadcast through the VCR (E-E operation), it is possible that the *video heads are dirty* (head gaps are slightly clogged). If there is sound but no picture, the video head gaps may be badly clogged. If the playback picture is unstable with a new TV set (never previously used with the VCR), it is possible that the AFC circuits of the TV are not compatible with the VCR. This problem is discussed in Sec. 2-7.1. If there is a color beat (rainbowlike stripes on the screen), the problem may be interference rather than a failure in the VCR or TV.

2-6.2 Operational checklist

The following checklist describes symptoms and possible causes for some basic VCR troubles.

Record button cannot be pressed. Check that there is a cassette installed and that the safety tab has not been removed from the cassette. If necessary, cover the safety tab hole with tape.

No E-E picture; no picture and sound that you wish to record. Check that the program select switch is in the correct position. Check the fine tuning on the TV inactive channel.

No color or very poor color. If there is no color on playback, check the fine tuning on the TV inactive channel. Note that if the VCR fine tuning is maladjusted during record, color may appear while recording, but may not appear during playback. Always check fine tuning of both the VCR and TV as a first step when there are color problems.

Playback picture is unstable. If you have periodic problems with picture instability, before tearing into the VCR with a pickax, check the following points. Has the VCR been operated in an area having a different ac line frequency? While recording, it is possible that a fringe-area signal was weak (intermittently) so that the sync signal was not properly recorded. During recording, there could have been some interference or large fluctuations in the power supply voltage. The cassette tape could be defective. The tracking control could be improperly adjusted.

Snow noise appears on the picture during playback only. Check the tracking control!

Sound but no picture, and excessive black-and-white snow noise. Check for very dirty video heads.

Upper part of picture is twisted or entire picture is unstable. The time constant of the AFC circuits in the TV is not compatible with the VCR. Refer to Sec. 2–7.1.

Tape stops during rewind. Is the memory counter switch on? If the memory switch is on, the tape stops automatically at 9999 or 0000 during rewind (on most VCRs).

The rewind and fast-forward buttons cannot be locked or operated. Is the cassette tape at either end of its travel? If the tape is at the beginning, rewind does not function. Fast forward does not function if the tape is at the end.

Cassette will not eject. Is the power on?

Acoustic feedback (whistlelike sound) when recording with camera and microphone. Keep the microphone away from the TV. Turn down the TV volume.

Noise band in playback picture, picture unstable, with too high- or too low-pitched sound. In some VCRs the tape is automatically locked to the correct speed by the servo. However, other VCRs also require some form of manual switching (to select SP, LP, and EP or SLP). Always check the operating controls for such a possibility, especially when you get sound that is constantly too high or too low in pitch.

2-7 VCR TROUBLESHOOTING/REPAIR NOTES

The following notes summarize practical suggestions for troubleshooting all types of VCRs. The notes are arranged by the major functional sections of a VCR (video, audio, servo, etc.) in which the troubles are most likely to occur.

2-7.1 TV AFC compatibility problems

If the AFC circuits of the TV are not compatible with the VCR, *skewing* may result. Generally, the term *skew* or *skewing* in VCRs is applied when the upper part of the reproduced picture is bent or distorted by *incorrect back-tension* on the tape. However, the same effect can be produced when the time constant of the TV AFC circuits cannot follow the VCR playback output. This condition is *very rare* in newer TV sets (which are designed with VCRs and videodisc players in mind), and appears only in about 1% of older TV sets (and almost never when the VCR and TV are produced by the same manufacturer). So, do not change the TV AFC circuits unless you are absolutely certain there is a problem. Try the VCR with a different TV. Then try the TV with a different VCR.

Once you are *certain* that there is a problem with compatibility, you can reduce the time constant of the *integrating circuit* associated with the TV AFC. Figure 2-9 shows the major components of a typical TV AFC integrating circuit. To reduce the time constant, reduce the values of either or both capacitors C1 and C2, reduce the value of R1, and increase the value of R2. It is generally not necessary to change all four values. Be sure to check the stability

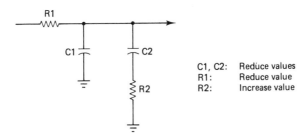

FIGURE 2-9 Altering the integrating circuits of a TV AFC

of the TV horizontal sync after changing any AFC value, since the AFC circuits are usually part of the horizontal sync system in modern TV sets.

2-7.2 Tuner and RF unit problems

When a VCR is first connected to a TV, it is likely that the unused channel (3 or 4) of the TV is not properly fine-tuned. When fine tuning the TV, operate the VCR in the *playback mode* using a known good cassette, preferably one with a color program. If you try to fine-tune the TV in the record or E-E mode, both the VCR and TV tuners are connected in the circuit and the picture is affected by either or both tuners. With playback, the picture depends only on the TV tuner. Once the normally unused channel of the TV is fine-tuned for the best picture, the VCR tuner can be fine-tuned as necessary.

Replacing the RF unit. In virtually all VCRs, the RF unit (also called the modulator or possibly the converter) must be replaced in the event of failure. No adjustment is possible, and internal parts cannot be replaced on an individual basis. This is because the RF unit is essentially a miniature TV transmitter and must be *type accepted* using very specialized test equipment, as are other transmitters. You must replace the RF unit as a package if you suspect failure. For example, if you have found proper audio and video inputs (and power source) to the RF unit, but there is no output (or low output) at the unused channel, the problem is likely to be in the RF unit. As a point of reference, a typical RF unit produces 1000 μV into a 75-Ω load (or 2000 μV into a 300-Ω load) on the selected channel. Signals outside the channel frequency by more than 3.5 MHz are reduced at least 35 dB. Also, there is a 60-dB drop introduced by the changeover switching network between the RF unit and antenna.

Replacing the tuner. In many VCRs the tuner is also replaced as a unit in the event of failure. However, some manufacturers supply replacement parts for their tuners. Also, most manufacturers provide for tuner adjustment as part of service, as discussed in Chapter 10. As a point of reference, a typical VCR tuner (including the IF) produces 1 V (p-p) of video into a 75-Ω load. Typically, audio output from the tuner is in the -10- to -20-dB range.

2-7.3 Copy problems

It is possible to copy a videocassette using two VCRs. One VCR plays the cassette to be copied, while the other VCR makes the copy. Keep two points in mind when making such copies. First, if the cassette being copied contains any copyrighted material, you may be doing something illegal! Second, a copy is never as good as the original, and copies of copies are usually terrible. Even with professional recording and copying equipment, the quality of a copy (particularly the color) deteriorates with each copying. The quality of a first copy (called the second generation) can be acceptable provided that the original is of very good quality. However, a second copy (third generation) is probably of unacceptable quality. Forget fourth generation (or beyond) copies. So if you are called in to service a VCR that "will not make good copies of other cassettes," explain that the problem probably has no cure.

2-7.4 Video camera sync and interlace problems

If you are to service a VCR that operates properly in all modes, except when used with a known good video camera, the problem may be one of incompatibility. The cameras recommended for use with a VCR (generally of the same manufacturer) should certainly produce good cassette recordings. In general, most cameras designed for use with VCRs, even though of a different manufacturer, are compatible with any VCR. Such cameras are usually designated as having a *2:1 interlace*. Essentially, a 2:1 interlace means that both the vertical and horizontal sync circuits of the camera are locked to the same frequency source (possibly the power line) by a definite ratio. When operating a camera, the sync signal normally supplied by the TV broadcast is obtained from the camera and recorded on the control track of the VCR tape.

Some inexpensive cameras, particularly those used in surveillance work, have a *random interlace* where the horizontal and vertical sync are not locked together. The playback of a recording made with a random interlace camera often has a strong *beat pattern* (*herringbone effect*). One way to confirm a random interlace condition is to watch the playback while observing the last horizontal line above the vertical blanking bar. Operate the TV vertical hold control as necessary to roll the picture so that the blanking bar is visible. If the end of the last horizontal line is stationary, the camera has 2:1 interlace and should be compatible. If the last horizontal line is moving on a camera playback, the camera is not providing the necessary sync and probably has random interlace.

2-7.5 Wow and flutter problems

VCRs are subject to wow and flutter, as are most audio recorders. Wow and flutter are tape-transport speed fluctuations that may cause a regularly occurring instability in the picture and a quivering or wavering effect in the sound during record and playback. The longer fluctuations (below about 3 Hz) are called *wow*; shorter fluctuations (typically 3 to 20 Hz) are called *flutter*. Wow and flutter can be caused by mechanical problems in the tape transport or by the servo system. Wow and flutter are almost always present in all VCRs, but it is only when they go beyond a certain tolerance that they are objectionable.

If you are to service a VCR where the complaint appears to be excessive wow and flutter, first check the actual amount. This is done using the low-frequency tone recorded on the alignment tape and a frequency counter connected to the audio line at some convenient point. Typically, the low-frequency tone is in the order of 333 Hz, and an acceptable tolerance is 0.03%. If necessary, operate the frequency counter in the period mode to increase resolution (Sec. 2-2.4).

2-7.6 System control problems

It is difficult to generalize about system control (Chapter 14) service problems. In most modern VCRs, system control is performed by a microprocessor/microcomputer (possibly several microprocessors interconnected by buses). In the simplest of terms, you press front-panel buttons to initiate a given operating mode, and the microprocessors produce the necessary control signals (to operate relays, motors, etc.). Each VCR has its own system control functions. You must learn these functions.

However, system control for all VCRs has certain *automatic stop* functions, such as end-of-tape stop, condensation detector (dew sensor), and so on, which must be accounted for in service. For example, when checking any system control, make certain that all the automatic stop functions are capable of working. Next, make certain that some automatic function has not worked at the wrong time (end-of-tape stop occurs in the middle of the tape). Equally important, make certain that a normal automatic stop is not the cause of your imaginary trouble. (Do not expect the tape to keep moving when the end-of-tape stop has occurred.) Also, it is very often necessary to disable the automatic stop function during service. The following notes describe some generalized procedures for checking and testing system control.

Slack tape sensors can be checked by visual inspection and by pressing on the tape with your fingers to simulate slack tape. If the slack tape sensor includes a microswitch (such as with some Beta VCRs), the sensor circuit can be disabled by forcing a match or cardboard against the sensor to keep the microswitch from triggering.

End-of-tape sensors can be checked by simulating the end-of-tape condition. For VHS, this involves exposing a photosensor to light (to simulate the clear plastic tape leader) to trigger automatic stop. For Beta, the end-of-tape foil can be simulated by placing a piece of foil near the surface of the sensing coil. To disable the VHS end-of-tape sensor, it is necessary to cover the phototransistor with opaque tape or a cap. Do not remove the light source for the end-of-tape sensor on a VHS machine! In most cases, removal of the sensor lamp is sensed by a failure circuit which also triggers auto stop. (Even with the phototransistor covered, stray light may trigger the auto stop condition.)

If the VCR has a switch that is actuated when a cassette is in place, such as a *cassette-in switch* on a top-load Beta or a *cassette-up switch* on a front-load VHS, locate the mechanism that actuates the switch and hold the mechanism in place with tape. In many (but not all) cases, it is possible to operate the VCR through all its modes without a cassette installed if the cassette switch can be actuated manually.

Take-up reel detectors (VHS) can be checked by holding the take-up reel. This causes the take-up reel clutch to slip (to prevent damage), but the detector senses that the reel is not turning and produces automatic stop.

Always check that all automatic stop functions work and that all bypasses and simulations are removed after any service work!

2-7.7 Interchange problems

When a VCR can play back its own recordings with good quality, but playback of tapes recorded on other machines is poor, the VCR is said to have *interchange* problems; it is unable to interchange tapes. Such problems are almost always located in the mechanical section of the VCR, usually in the tape path. Quite often, interchange problems are the result of improper adjustment. Manufacturers sometimes include interchange adjustments as part of their overall electrical/mechanical adjustments.

The simplest way to make interchange adjustments is to monitor the RF output from the video heads during playback and adjust elements of the tape path to produce a maximum, uniform RF output from a factory alignment tape. Generally, the output is measured at a point after head switching so that all heads are monitored. Always follow the manufacturer's adjustment procedures.

2-7.8 Servo system problems

Total failures in the servo system (Chapters 7 and 8) are usually easy to find. If a servo motor fails to operate, check that the power is applied to the motor at the appropriate time. If power is there but the motor does not operate, the motor is at fault (burned out or open windings, etc.). If the power is absent, track the power line back to the source. (Is the microprocessor delivering the necessary control signal or power to the relay or IC?)

The problem is not so easy to locate when the servo fails to lock on either (or both) record and playback, or locks up at the wrong time (causing the heads to mistrack even slightly. Obviously, if the control signal is not recorded (or is improperly recorded) on the control track during record, the servo cannot lock properly during playback. Therefore, your first step is to see if the servo can play back a properly recorded tape. There are several ways to do this.

There are usually some obvious symptoms when the servo is not locking properly. (There is a horizontal band of noise that moves vertically through the picture if the servo is out of sync during playback.) The picture may appear normal at times, possibly leading you to think that you have an intermittent condition. However, with a true out-of-sync condition, the noise band appears regularly, sometimes covering the entire picture.

Keep in mind that the out-of-sync condition during playback can be the result of servo failure or the fact that the sync signals (control signals) are not properly recorded on the tape control track. To find out if the servo is capable of locking properly, play back a known good tape. If the playback is out of sync you definitely have a servo problem.

The symptoms for failure of the servo to lock during record are about the same as during playback, with one major exception. During record, the head switching point (which appears as a break in the horizontal noise band) appears to move vertically through the picture in a random fashion.

Another way to check if the servo is locking on either record or playback involves looking at some point on the rotating scanner or video head assembly under fluorescent light. When the servo is locked, the fluorescent light produces a blurred pattern on the rotating scanner that appears almost stationary. When the servo is unlocked, the pattern appears to spin. Try checking the scanner of a known good VCR under fluorescent light. Stop and start the VCR in the record mode. Note that the blurred pattern appears to spin when the scanner first starts, but settles down to almost stationary when the servo locks. Repeat this several times until you become familiar with the appearance of a locked and unlocked servo under fluorescent light.

Once you have studied the symptoms and checked the servo playback with a known good tape, you can use the results to localize the trouble in the servo. For example, if the servo remains locked during playback of a known good tape, it is reasonable to assume that the circuits between the control head and servo motors are good. In early-model VCRs such circuits include the playback amplifiers, tracking delay network, sample-and-hold circuits, power amplifier, servo motors and breaks, feedback pulses, and ramp generators. In modern VCRs the servo circuits include both speed and phase control circuits, where it is generally more difficult to localize trouble.

Keep in mind that servo troubles may be mechanical or electrical and may be the result of either improper adjustment or component failure, or both. As a general guideline, if you suspect a servo problem, start by making the electrical adjustments that apply to the servo. Always follow the manufacturer's adjustment procedures, using the circuit descriptions of Chapters 7 and 8 as guides. This may cure the servo problem. If not, following the adjustment procedure tells you (at the very least) if all the servo control signals are available at the appropriate points in the circuits. (Such control signals include all of the cylinder and capstan control pulses shown in Fig. 1-6.) If one or more of the signals are found to be missing or abnormal during adjustment, you have an excellent starting point for troubleshooting.

There are two points sometimes overlooked when troubleshooting a servo that fails to lock up. First, the free-running speed of the servo may be so far from normal that the servo simply cannot lock up. This problem usually shows up during adjustment. Second, on those VCRs that use rubber belts to drive servo motors (typically the early-model units rather than modern VCRs), the rubber may have stretched (or been otherwise damaged). If you have replacement belts available, compare the used VCR belts for size and conformation. Hold a new and used belt on your finger under no strain. If the used belt is larger or does not conform to the new belt, install the new belt and recheck the servo for proper lockup.

3

POWER SUPPLY CIRCUITS

This chapter is devoted to troubleshooting and repair of VCR power supply circuits. The chapter starts with descriptions of some typical circuits (a modern four-head VCR and a five-head convertible VCR) and concludes with troubleshooting/repair procedures for the specific circuits.

3-1 FOUR-HEAD VCR POWER SUPPLY

Figure 3-1 shows the power distribution in a typical four-head VCR (that shown in Fig. 1-1a). The circuit develops a total of nine dc voltages, plus 6-V ac, and consists of T971 with five secondary windings, three full-wave bridge rectifiers, two half-wave rectifiers, and six voltage regulators with their respective drive circuitry. The power supply is protected by four fuses and a fusible resistor. The primary winding is fused with a 1.6-A fuse; secondary windings 1, 2, and 4 are each individually fused; and winding 3 is protected by a fusible resistor R903.

Secondary winding 1 supplies ac to D901, which outputs 15 V to the audio, operation (system control), and timer processor circuits. Four additional voltages are taken from the 15-V source. Standby 9V is taken via standby 9-V regulator Q971, and the 7-V on-switch Q912-Q913 (Sec. 3-1.2). The standby 9 V is applied to the servo, timer, and operations circuits via D914, and to the backup 9-V line via D915. If power is lost, the backup 9-V line is supplied from a backup power supply (part of the timer processor circuit de-

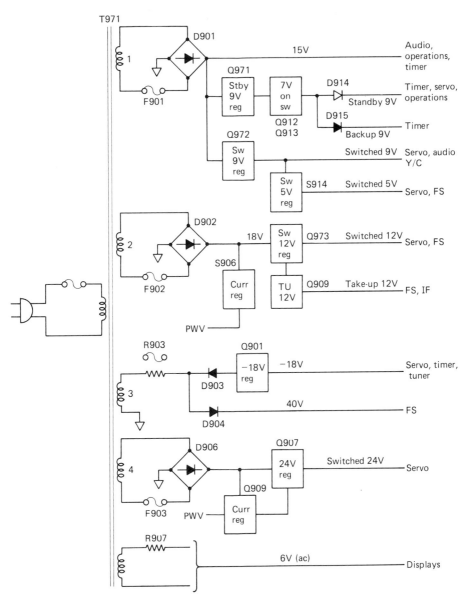

FIGURE 3-1 Power distribution in a typical four-head VCR

scribed in Chapter 13), and memory is maintained for approximately 30 minutes.

Two switched supplies (Sec. 3-1.1) are also taken from the 15-V supply. When the VCR is switched on, the timer microprocessor IC outputs a high on the PWV line (the "on" command line). This enables the switched 9-V regulator Q972 and supplies switched 9 V to the servo, audio, and Y/C circuits. The presence of switched 9 V also activates the switched 5-V regulator Q914, and 5 V is directed to the servo and FS circuits.

Bridge rectifier D902 develops 18 V used to supply the switched 12 V and TU (takeup) 12-V supplies. When the PWV line is high (VCR ON), Q906 is enabled, activating Q909 and in turn enabling Q973. Switched 12 V is directed to the servo and FS circuits; TU 12 V is applied to the FS (frequency synthesis) and IF circuits.

Secondary winding 3 of T971 directs ac to two half-wave rectifiers via fusible resistor R903. Rectifier D903 develops -18 V which is applied to the servo, timer, and tuner circuits. D904 develops 40 V, which serves as the FS system tuning voltage supply.

The switched 24-V supply for the servo circuits is derived from secondary winding 4 and bridge rectifier D906. A high on the PWV line enables the 24-V current regulator Q909 which, in turn, activates the switched 24-V regulator Q907. The remaining secondary winding 5 supplies 6-V ac for the digital display filaments via current limiting resistor R907.

3–1.1 Switched B+ supplies

Figure 3-2 shows the switched B+ circuits. Note that the switched supplies operate only when the VCR is switched on and are controlled by logic on the PWV line from the timer microprocessor (PWV is high when the VCR is on). When PWV goes high, the zener point of D924 is exceeded and a high is applied to Q910, turning Q910 on. The resulting low at the collector of Q910 turns on the five switched supplies.

The low from the collector of Q910 is inverted to a high by Q911. This high enables four of the switched supplies. The high reverse-biases D920, removing the low at pin 3 of IC901, enabling IC901. This activates Q972, which supplies 9 V to the switched 9-V line. The presence of switched 9 V turns on Q914, which supplies power to the switched 5-V line.

The high from Q911, in conjunction with the low at the collector of Q910, turns on the switched 12-V circuits. The low from Q910 is directed, via D917 and R916, to the base of Q906, forward-biasing Q906. At the same time, the high from Q911 reverse-biases D923, allowing the collector of Q906 to go high. The collector of Q906 is set to 12.6 V by zener D911 and is applied to the base of Q909. The conduction of Q909 serves two purposes: (1) It supplies 12 V to the TU 12-V line, and (2) it enables Q973, supplying 12 V to the switched 12-V line.

The switched 24 V is turned on by the low from Q910. The low turns on Q908 which, in turn, turns on Q907, supplying 24 V to the VCR circuits.

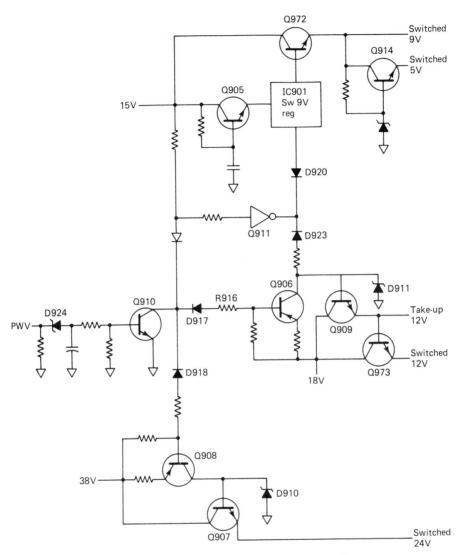

FIGURE 3-2 Typical switched B+ circuits

3-1.2 7-V on switch

Figure 3-3 shows the 7-V on-switch circuits. To ensure correction operation of the various microprocessors used in the VCR, the microprocessors must be initialized (reset) when power is applied to the VCR, setting all output to zero. The microprocessor power supply voltage must rise sharply when power is applied in order to produce correct initialization. The 7-V on-switch circuits

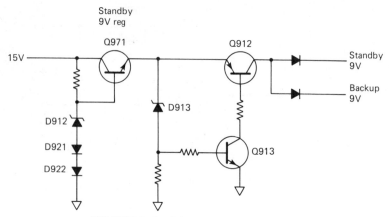

FIGURE 3-3 7-V on-switch circuits

provide this sharp rise in voltage on the standby 9-V and backup 9-V lines required in the initializing process.

When power is first applied to the VCR, or when power is restored after a power failure, the output of the standby 9-V regulator Q971 starts to increase and is applied to Q912. However, the base of Q912 is essentially "floating" since D913 is not conducting, so Q912 does not conduct. When the output of Q971 exceeds 7 V, zener D913 conducts, lowering the base of Q912 and turning Q912 on. With Q912 on, the output of Q971 is applied to the standby 9-V and backup 9-V lines.

Since Q912 does not conduct until the output of Q971 exceeds 7 V, the standby 9-V and backup 9-V lines have a sharp rise in voltage (0 to 7 V) when power is connected to the VCR. The microprocessors are then properly initialized to prevent erroneous operation.

3-2 FIVE-HEAD VCR POWER SUPPLY

Figure 3-4 shows the overall functions of the power supply for a five-head convertible VCR (that shown in Fig. 1-1c). Note that the convertible VCR consists of two major sections: (1) a portable, battery-operated VCR deck, and (2) a stationary tuner/timer/charger. When the sections are combined, the unit has both the appearance and features of a table model VCR. However, the portable VCR deck can be removed and operated with a camera and batteries. So the power supply circuits must provide the dual function of supplying power and charging the batteries.

As shown in Fig. 3-4, the power supply (within the tuner/timer/charger) provides a regulated 12.5 V, or 18 V, to a rear external battery jack. The regulator output is routed through a switch (within the jack) to a 26-pin receptacle (to supply power to the VCR deck). In the VCR mode, the regulator

Five-Head VCR Power Supply 91

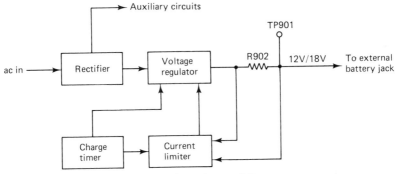

FIGURE 3-4 Overall functions of the power supply

output is regulated to 12.5 V and current limited to a maximum of 4 A. In battery charge mode, the output is regulated to 18 V and current limited to 1.3 A.

3-2.1 Rectifier circuit operation

Figure 3-5 shows the power supply rectifier circuits. The 14-V ac, routed through F951, is rectified by D951 and filtered by C952 to provide 18 V (unregulated) to the 12-V regulator circuit. Discharge switch S951 is used to discharge C952 when the rectifier circuit board is removed. Discharging C952 eliminates the possibility of damaging components when the rectifier board is reconnected.

The 7-V ac is rectified by D955 and regulated by IC951. The 5-V regulator output is increased to 5.6 V by the addition of D958, which elevates the ground terminal by 0.6 V. The 5.6-V output from IC951 is passed through diodes to isolate the various circuits and reduce the 0.6 V so as to again get the proper 5 V required by the VCR circuits.

The 5.6 V is applied through D910 to the remote control circuits through D51 to the tuner/timer/charger, through D52 to the timer backup circuits, and through D714 to the PLL in the tuner. Note that the 5 V from D52 also charges C51, which provides power for timer memory whenever ac power is lost. Capacitor C51 maintains power to the timer microprocessor for approximately one hour.

The 39-V ac routed through F952 is rectified to provide both positive and negative supply voltages. The positive portion is rectified by D956 and supplies 40 V to the regulator circuit (for base biasing of the regulator), and 40 V to the PLL circuit (for a tuning voltage). The negative portion of the 39-V ac is rectified by D957 and applied to the input of two regulators. One regulator, Q951-Q952, generates a −27 V for the timer display circuits. The second regulator, ZD951, generates a −3 V, which is applied to the 12-V regulator current-limiting section (Sec. 3-2.3).

92 Power Supply Circuits

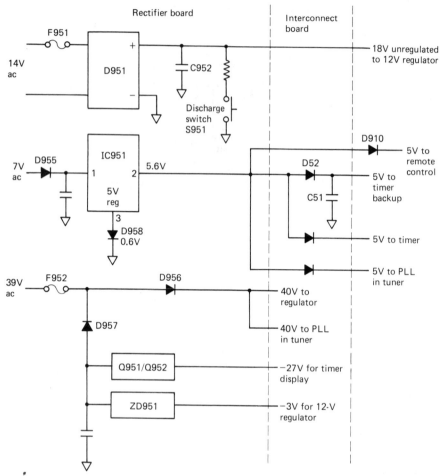

FIGURE 3-5 Power-supply rectifier circuits

Rectifier circuit troubleshooting/repair. First, if removal of the rectifier circuit board is required, remove ac power, press discharge switch S951, and then remove the circuit board. Please do not press S951 with the power applied.

During service of the tuner/timer/charger or the VCR deck, the output levels of the rectifier circuit should be checked. If any of the outputs are absent or abnormal, check the associated filter capacitor, transistor, and diodes. There are no adjustments in the rectifier circuits. Repair the defective circuit area before continuing on with troubleshooting.

Keep in mind that if one of the power supply output lines is shorted (partially or fully), this will affect the output voltage. Likewise, such a short can damage the rectifier circuits. So always look for shorts on the lines if you find any power supply output that is absent or abnormal (very low).

3-2.2 Voltage regulator and select operation

Figure 3-6 shows the voltage regulator and select circuits. The voltage regulator is switched between two different output voltages: 12.5 V during normal VCR operation (no charge), and 18 V during battery charge operation. The voltage regulator consists of series regulator Q901 and reference amplifier Q903. Feedback of the output voltage to Q903 is taken from the divided-down output at the junction of R912 and R918. R912 sets the output voltage level. The emitter of Q903 is biased by 7.5-V zener ZD901. Variations of the output

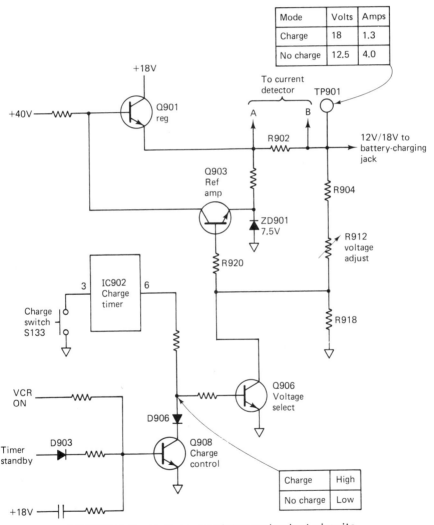

FIGURE 3-6 Voltage regulator and select circuits

voltage are passed to Q903, which in turn causes conduction changes in Q901 and Q903 to maintain a constant output voltage (at 12.5 V in the VCR mode).

In the charge mode, an additional resistor R920 is connected across R918 to change the resistor divider network, shifting the output to 18 V. Resistor R920 is switched into the circuit by Q906, which is turned on during the charge mode by operation of IC902. During charge, the output at IC902-6 is high, applying a forward bias to Q906. While in charge (battery only) the VCR can still be turned on manually, or the timer can be placed in the standby mode. In either of these modes, the 18-V output must be switched to 12.5 V to prevent damage to the other circuits.

To switch the output to 12.5 V during charge, the VCR-on signal from the VCR or the timer standby signal from the timer microprocessor is applied to Q908. This turns Q908 on, causing D906 to conduct, and pulls the base of Q906 low. Q906 turns off, disconnecting R920, and the output is restored to 12.5 V. Also, during initial power-up of the VCR, the unregulated 18 V is passed through a capacitor/resistor network to Q908 to assure that the regulator applies 12.5 V to the circuits.

The output current of the 12-V regulator passes through R902. The voltage developed across R902 is passed to the current limit circuit (Sec. 3-2.3). By monitoring the voltage across R902, the current limit circuit sets the maximum output to 1.3 A in charge or 4 A in normal (VCR) mode.

Voltage regulator troubleshooting/repair. The three most common symptoms of voltage regulator problems are low output voltage, excessive output voltage, and no charging operation.

If the voltage is low, try correcting the condition by adjustment of R912. Monitor TP901 and adjust R912 for 12.5 V (in the normal VCR mode). If this does not correct the problem, look for 7.5 V at the emitter of Q903. If low, ZD901 may be leaking. If Q903 is at 7.5 V, suspect a defective Q903 or Q901. Also check the current limiter circuits (Sec. 3-2.3).

If the voltage is high, try correcting the problem by adjustment of R912, and then check the emitter of Q903. If high, ZD901 may be open. Measure the base voltage of Q903 (about 8V). If low, suspect an open R904 or R912. Measure the base voltage of Q901. In the normal VCR mode, if the base of Q901 is greater than about 13.5 V, suspect a leaky Q901 or open Q903.

If there is no charging operation, check for 18 V at TP901 in the battery charge mode. This can be done by connecting a 120-Ω, 5-W resistor from TP901 to ground to simulate the load of a discharged battery, and pressing the charge switch (connected to the charge timer IC902). Confirm a high output at IC902, pin 6, after the charge button is pressed. If the high is missing, refer to battery charge circuit troubleshooting (Sec. 3-2.4). If the high is present at IC902-6, check for a high at the collector of Q908. If the collector of Q908 is low, suspect a defective Q908. If the collector Q908 is high, check for a low dc

Five-Head VCR Power Supply   95

voltage on the collector of Q906. If the collector of Q906 is high, suspect an open Q906.

3-2.3 Current limit select operation

Figure 3-7 shows the current limit select circuits. The power supply is current limited to 1.3 A during battery charge and about 4 A in normal VCR mode, as determined by the current through R902. The voltage developed across R902 is passed to the input of op-amp IC902. The noninverting input (5) is held constant by a fixed bias, while the inverting input (6) varies in accordance with the current through R902. If the current goes above the desired limits, the

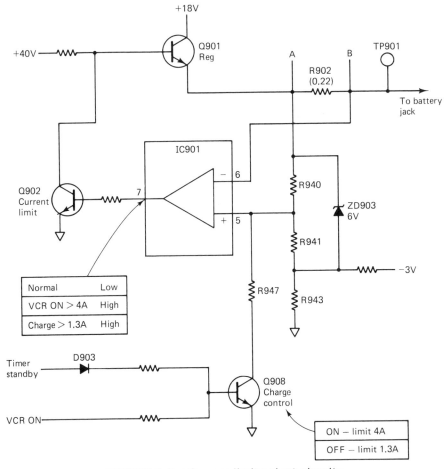

FIGURE 3-7 Current-limit select circuits

output of IC901 (pin 7) goes high, turning on Q902 and reducing the base voltage of Q901. This reduces both the output voltage and current.

The current required to switch the output of IC901 to high is set by the voltage across R902 and the fixed voltage applied to IC901-5 by the divider network. During battery charge, Q908 is off, R947 is out of the circuit, and the fixed bias to IC901-5 is such as to limit the current at 1.3 A. When in the timer-on or VCR-on modes, Q908 is turned on, R947 is connected across R941/R943, and the bias to IC901-5 is altered as necessary to limit the current at about 4 A.

Current limit troubleshooting/repair. To confirm proper operation of the current limit circuit, connect a *fused* ammeter from TP901 to ground. The ammeter must be capable of reading a minimum of 5 A. In the normal VCR mode, confirm that the maximum current drawn by the meter is about 4 A. In the battery charge mode, confirm that the output is about 1.3 A.

If either of the currents are abnormal, look for an increased or decreased resistance in R902 (it should be 0.22Ω) or defective Q908/IC901.

If you cannot get the proper output voltage during normal VCR mode, check that Q902 is not on by checking for a low at IC901-7. If IC901-7 is high (during normal VCR mode) suspect a defective IC901 or R902.

3-2.4 Battery charge circuit operation

Figure 3-8 shows the battery charge circuits. The battery is nickel cadmium (ni-cad), and has three terminals as shown. One terminal (+) is connected to the positive side of the power supply, while the second (−) terminal is connected to the negative output. During VCR operation, the battery current flows through an 80°C thermostat (circuit breaker) located within the battery case. The 80°C thermostat monitors the internal temperature of the battery. In case of a shorted load, the internal battery temperature rises, opening the thermostat to disconnect the battery from the circuit, thus preventing damage to the battery.

The third (T) terminal is connected to the negative side of the battery through a 45°C thermostat. When the battery reaches *full charge,* the current drawn by the battery causes the internal temperature to rise. When the internal temperature of the battery reaches 45°C, the 45°C thermostat opens, signaling the charging circuit that full charge has been reached. The opening of the 45°C thermostat ends the charge cycle and shifts the output of the regulator to 12.5 V.

When the charge switch (S133, located on the tuner/timer/charge front panel) is pressed, IC902 generates a high at pin 6. This high is applied to IC902-2 (and to the base of Q906, Fig. 3-6), turning on the oscillator portion of IC902. The high is also applied to the base of Q921, turning on Q921.

When Q906 is turned on (Fig. 3-6), the voltage regulator output goes to

Five-Head VCR Power Supply   97

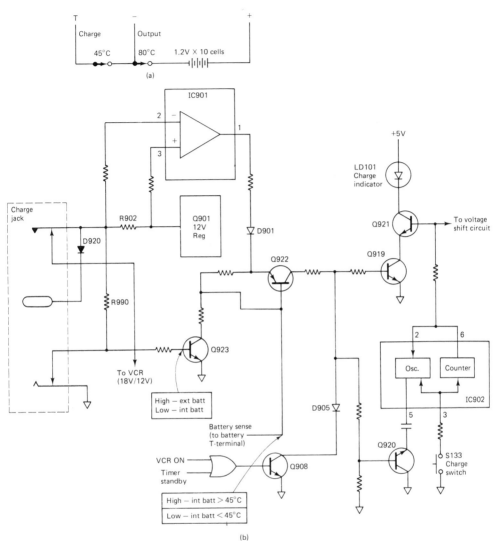

FIGURE 3-8 Battery charge circuits

18 V (as discussed in Sec. 3-2.2). When the 18 V is applied to the battery, the battery current is increased to the current limit range of 1.3 A. As discussed in Sec. 3-2.3, the charging current is detected by R902 and IC901. During charge, the output of IC901 (pin 1) goes high, pulling the emitter of Q922 high.

The battery sense line (connected to the battery T terminal) is grounded through the 45°C thermostat, keeping the base of Q922 low. During charge, when the emitter of Q922 is high, the low at the base turns on Q922, allowing the collector to go high. The high at the Q922 collector turns on Q919, turning on charge indicator LD101 (through Q919/Q921). The Q922 collector high also turns on Q920, which acts as a switch to ground one side of a capacitor connected to pin 5 of IC902. When Q920 is turned on, the capacitor is connected from IC902-5 to ground. This causes the oscillator within IC902 to operate at 4 Hz. The low frequency is divided down by an internal counter within IC902 to eventually pull pin 6 low after the counter times out. When IC902-6 goes low, the charging cycle ends as the regulator output voltage drops to 12.5 V. It takes about 70 minutes for the counter to time out.

The charge cycle is terminated in one of three ways. First, the 45°C thermostat opens, causing the base of Q922 to go high. This turns off Q922 and turns on Q920, removing the capacitor from IC902-5. Without the capacitor, the frequency changes from 4 Hz to 60 kHz, causing IC902-6 to go low immediately and turning off the voltage select circuits (Fig. 3-6) to resume normal 12.5-V operation of the regulator.

Second, during charge, if either VCR-on or timer-on signals are applied to Q908, the base of Q920 goes low, disconnecting the capacitor from IC902-5 and thus causing the counter to time out.

Third, the charge cycle is ended after about 70 minutes due to normal time-out of the counter.

During charge of an external battery connected at the rear of the tuner assembly, an accessory adapter is used so that the three-terminal battery is converted to a standard two-conductor cable and plug assembly. The plug diameter is different from that of lead-acid batteries used in some other VCRs. This is done to assure that the two types of batteries are not interchanged, because the charging systems are not compatible.

The adapter connects the T terminal of the ni-cad battery to the negative side of the two-conductor plug. When this charge plug is connected to the rear of the tuner, an internal switch is activated, disconnecting power to the VCR as shown in Fig. 3-8, so the VCR *cannot be operated* during charge of an *external battery*. A second switch is also activated, removing the ground from the base of Q923. Without the ground, the base of Q923 is pulled high through R990, turning on Q923 and pulling the base of Q922 low. When the charge switch S133 is pressed, the external battery draws current and IC901 turns on, applying a high at IC901-1. The remaining operation for charging an external battery is similar to that for charging the internal battery.

During the external battery charge mode, the charge cycle is terminated in one of two ways. Either the internal 45°C thermostat opens, or the normal 70-minute charge time ends.

Battery charge troubleshooting/repair. If the battery charge circuit does not operate properly, disconnect the VCR and external battery from the tuner/timer/charger. Connect a 120-5-W resistor from the anode of D920 (Fig. 3-8) to ground. Press the charge switch S133 while monitoring the voltage at

IC902-6. The output at pin 6 of IC902 should go high. If it does not, suspect IC902 or S133.

If IC902-6 goes high when S133 is pressed, check for 18 V at TP901 (Fig. 3-7). If the 18 V is absent or abnormal, check the voltage select as discussed in Sec. 3-2.2. If the voltage at TP901 is correct (18 V), check for a high at IC901-1 (Fig. 3-8). If the high is missing, suspect IC901. If the high is present at IC901-1, check for a high at the collector of Q922. If normal, check for a high at the anode of D905. If the high is missing, suspect Q922 or Q923. If the anode of D905 is low, suspect Q908 (or check if VCR-on and timer signals are applied through the OR gate to the base of Q908). If the anode of D905 is high, but there is no battery charge operation, suspect Q919, Q920, IC902, and the capacitor at IC902-5. Note that if only Q919 is defective, it is possible for the battery charge operation to be good, but the charge indicator LD101 may not turn on.

REMOTE-CONTROL CIRCUITS

This chapter is devoted to troubleshooting and repair of VCR remote-control circuits. Most modern VCRs use some form of *infrared* (IR) remote control similar to that used in modern TV sets. Likewise, most VCR remote-control systems use some form of *position modulation* (P–M). So we concentrate on both IR and P–M remote control in this book. The chapter starts with descriptions of some typical remote-control transmission formats, and then goes on to cover some specific remote-control circuits. The chapter concludes with troubleshooting/repair procedures for some typical remote-control circuits.

4-1 TYPICAL REMOTE-CONTROL TRANSMISSION FORMATS

Figure 4–1 shows the transmission format of an IR remote-control system for a typical VCR (that shown in Fig. 1–1a). The system uses the P–M technique. The basic oscillator frequency of the hand transmitter is 455 kHz, generating a 38-kHz carrier for the transmitted signal, as shown in Fig. 4–1, waveform 1. The 38-kHz carrier is transmitted in bursts of 10 cycles each, with a burst width of 264 μs. The distance (or time) between the *leading edges* of succeeding bursts determines the digital logic (binary code) state represented by the signal.

The code for a digital logic low (0) is illustrated in Fig. 4–1, waveform 2. As shown, the time between the leading edges of the 38-kHz bursts is 1.05 ms for a digital low (0). A digital high (1) is indicated in waveform 3 of Fig.

Typical Remote-Control Transmission Formats 101

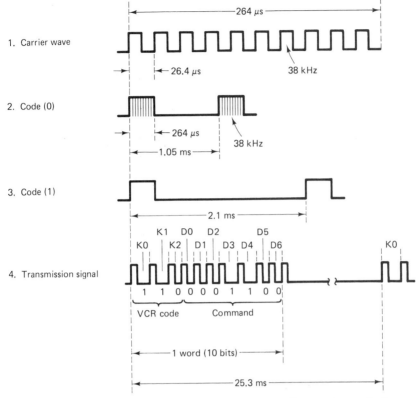

FIGURE 4-1 Transmission format of an IR remote-control system

4-1, and is represented by a 2.1-ms distance between the leading edges of succeeding bursts.

The transmission signal consists of a 10-bit *digital word*, as shown in Fig. 4-1, waveform 4. The total duration of one word is 25.3 ms. The first three bits of the transmitted word (K0,K1,K2) are always digital 110, which is the identification code for the particular VCR. The use of identification codes prevents interaction between IR remotes used on various types of equipment (between TVs, VCRs, and videodisc players, for example).

The remaining 7 bits (D0 through D6) designate the specific remote command. The digital 0001100 code illustrated in waveform 4 of Fig. 4-1 designates the STOP command. Note that this particular code for a STOP command is unique to this VCR. The command is meaningless (hopefully) to any other VCR (or TV), even those using a similar format but a different identification code.

4-2 TYPICAL REMOTE-CONTROL TRANSMITTER

Figure 4-2 shows the circuits of a typical hand-held IR remote transmitter. The circuits shown are unique in one respect. The reference frequency of the transmitter is 255 kHz (rather than the typical 455 kHz). This variation in transmission frequency prevents the transmitter from interfering with remote operation of TVs, other VCRs, and so on, and vice versa. The digital code representing a given function is represented by a series of "bursts" from the transmitter. Each burst contains 10 pulses, of 50 μs duration each, at a frequency of 20 kHz (rather than the typical 38 kHz). The duration of each burst is always 500 μs. The distance between recurring bursts identifies the 1s and 0s comprising the digital code. In this case, distance is defined as the amount of time existing between the rise of the first burst and the rise of the succeeding burst, as shown in Fig. 4-3. A logic 0 is 2 ms, while a logic 1 is 4 ms.

As shown in Fig. 4-2, the transmitter contains a single IC (ICM01), five transistors, a keyboard, and internal batteries. The function of ICM01 is to: (1) generate an oscillator signal for the creation of scanner signal outputs, which are applied to specific inputs of the IC (through the keyboard) to identify specific functions, and (2) to decode the information received at the inputs and produce a digital code representing that function.

No scanner output is available unless a transmitter key is pressed. When a given key is pressed, the oscillator in ICM01 is turned on and generates the appropriate scanner signals. The scanner outputs at pins 4, 5, 6, 7, and 8 of ICM01 are connected to one contact of specific keyboard switches. The remaining contact of each switch is connected to a specific input at pins 11, 12, 13, 14, and 15 of ICM01.

When a given key is pressed (enabling the oscillator), a scanner signal is applied to the corresponding input through the closed switch contacts. A specific scanner signal is applied to a specific input for each switch closure. Likewise, each keyboard switch is represented by a unique combination of scanner-output/signal-input. The scanner signal applied to a particular input is decoded by ICM01 and produces a specific binary code representing the function of the corresponding keyboard switch.

The binary code (representing the selected function) at pin 17 of ICM01 is applied to the base of QM01 and QM07. The code signal is amplified by QM01 and QM02, delayed for 2.2 ms, differentiated, and applied to QM03, which conducts for 550 μs, enabling the 40-kHz multivibrator QM05. The output of QM05 is applied to the base of QM07, along with the output from ICM01-17. The combination of both signals turns on QM07, activating the IR diode DM02. (Since two signals are required to activate DM02, erroneous transmission is prevented.) Diode DM02 is pulsed and the IR output is sent to the light-sensitive receiver diode on the VCR. The output from DM02 consists of a series of bursts (Fig. 4-3), with the spacing between bursts determined by the binary code being transmitted.

Typical Remote-Control Transmitter 103

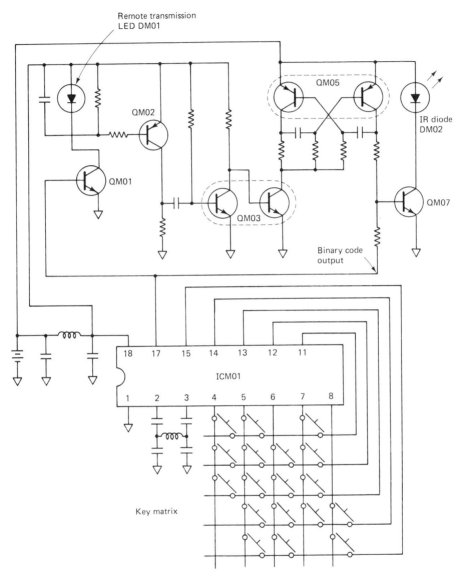

FIGURE 4-2 Typical hand-held IR remote-transmitter circuits

Remote-control transmitter troubleshooting/repair. Once you have definitely pinpointed the problem to a remote-control transmitter by trying a different transmitter with the same VCR, look for weak or defective batteries in the transmitter. Batteries are the most common causes of trouble in any

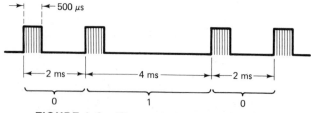

FIGURE 4-3 IR remote-transmitter pulses

remote-control transmitter. So start troubleshooting by putting in a known good battery. The next most common cause of trouble is a defective switch contact. Defective switches usually show up when one or more functions are absent or abnormal, but other functions are normal. (Say that you cannot select record, but you can select playback, with the remote transmitter.)

Note that when QM01 is turned on by the binary code output from ICM01-17, the remote transmission LED DM01 on the transmitter is turned on. This indicates that the transmitter is sending commands to the receiver diode on the VCR. (Of course, it is possible that the transmitter is operating properly to control the VCR, and only DM01 is defective. However, this is not likely.) So, if DM01 does not turn on when a key is pressed, look for a defective battery. If the battery is definitely good, suspect ICM01 or QM01.

If the batteries and switches appear to be good, but you cannot transmit any commands with the remote transmitter, press the keys while monitoring pin 17 of ICM01 with an oscilloscope or logic probe. Although it is usually not practical to determine the actual code being transmitted, the presence of pulse bursts usually indicates that ICM01 is good.

Next, try monitoring the pulse bursts through from QM01 to DM02. Also look for 40-kHz square waves at the collectors of QM05. If pulse bursts appear at DM02, but the remote transmitter cannot transmit commands, suspect DM02.

4-3 TYPICAL REMOTE-CONTROL PREAMPLIFIER AND NOISE FILTER

Figure 4-4 shows the remote-control preamplifier and noise filter circuits for a typical VCR (that shown in Fig. 1-1a). The transmitted IR signal from the remote-control transmitter is received by light-sensitive diode D511, which converts the signal from optical to electronic. The converted signal is applied to pin 1 of preamp IC502, where the signal is amplified, limited, and inverted. The output from IC502 at pin 16 represents the same binary code originally transmitted (such as shown in Fig. 4-1).

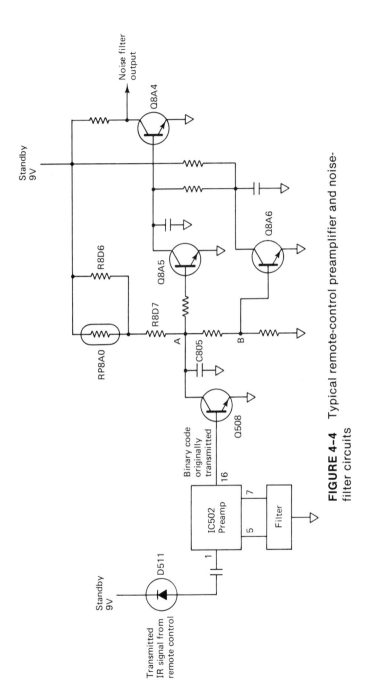

FIGURE 4-4 Typical remote-control preamplifier and noise-filter circuits

The output from IC502 is applied to a noise filter through Q508. The primary purpose of the noise-filter circuit of Fig. 4-4 is the prevention of erroneous remote triggering due to high-amplitude, long-duration, spurious noise pulses picked up along the path of the remote signal. The inverted remote signal from IC502 is applied to the base of Q508. Since the remote signal is inverted, the periods of the negative-going pulses are 264 μs. Q508 conducts only during the positive-going pulse periods, producing a low at the collector of Q508, a low at the base of Q8A5, a high at the collector of Q8A5, and a low at the collector of Q8A4, which is the noise-filter output.

At the fall of the input pulse, Q508 is disabled and remains so for the duration of the negative-going input pulse (264 μs). However, the collector of Q508 does not go high instantly due to the integrating action of R8D6/RP8A0, R8D7, and C805. Instead, the collector voltage at Q508 rises gradually as C805 charges, producing a sawtooth voltage as shown in Fig. 4-5, waveform 2A. When the sawtooth voltage reaches the threshold level, Q8A5 conducts, disabling Q8A4 and producing a high output to the circuits controlled by the remote circuit.

Due to the time constant of C805, R8D7, R8D6/RP8A0, an input pulse (or noise pulse) of a duration less than 200 μs does not allow sufficient time for the charge across C805 to reach the turn-on threshold of Q8A5. So the output of Q8A4 remains low for noise pulses. The output from Q8A4 goes

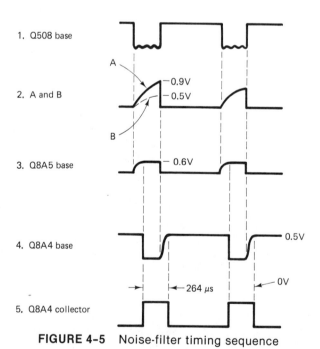

FIGURE 4-5 Noise-filter timing sequence

high only when an input pulse exceeds 200 μs, as is the case with the 264-μs pulses comprising a remote transmission.

Should a noise pulse exceed 700 μs in duration, Q8A6 comes into action. A normal incoming pulse period of 264 μs is too short for the C805 charge voltage (Fig. 4-5, waveform 2B) to reach the threshold level of Q8A6. As a result, Q8A6 does not normally conduct. However, a noise pulse of 700 μs allows the off-time of Q508 to increase sufficiently to allow the charge across C805 to exceed the threshold of Q8A6, turning on Q8A6 and driving the base of Q8A4 low.

When the noise pulse recedes, Q8A6 turns off. However, due to the integrating action of C8A7 and R8E3, the collector voltage of Q8A6 rises gradually, holding Q8A4 off until the turn-on threshold is reached (a duration of about 30 ms). The effect of this action is an interruption of the remote signal in the event of noise. If the noise is gone, the next transmitted code sequence is free to pass through the noise-filter circuit unaltered. The result is a faithful, noise-free reproduction of the transmitted remote signal at the collector of Q8A4.

4-4 TYPICAL REMOTE-CONTROL DECODER MICROPROCESSOR

Figure 4-6 shows the remote-control decoder microprocessor circuits for a typical VCR (that shown in Fig. 1-1c). The remote-control microprocessor IC401 has the dual function of decoding digital commands from the remote transmitter and converting the commands into signals (that are applied to the VCR system control, tuner, and power circuits, etc.). As shown, IC401 controls a multitude of VCR functions. The control instruction or command to IC401 is taken from the signal transmitted by the remote-control transmitter (Sec. 4-2). The transmitted IR signal is received by light-sensitive diode PD497 and converted into an electrical signal which is applied to amplifier/detector IC491. The detected signal from IC491-1 is amplified and inverted by Q401 and applied to the input of IC401-12. (This signal can be monitored with an oscilloscope or logic probe at IC401-12.)

The signal received from the remote transmitter is a 32-bit digital signal very similar in appearance to the signal used in the RCA projection TVs. (The logic signals have the same specification, or time periods, as the projection TV, but the digital code is different.) The code transmitted as the first 16 bits of the signal indicates to IC401 that the signal is for VCR operation. IC401 decodes the second 16 bits to determine which function has been requested, and turns on the corresponding output or outputs.

The output at IC401-36 is the power-on signal, which turns on transistor Q405 whenever the power button on the remote control transmitter is pressed. Transistor Q405 is in parallel with the manual on/off switch of the VCR,

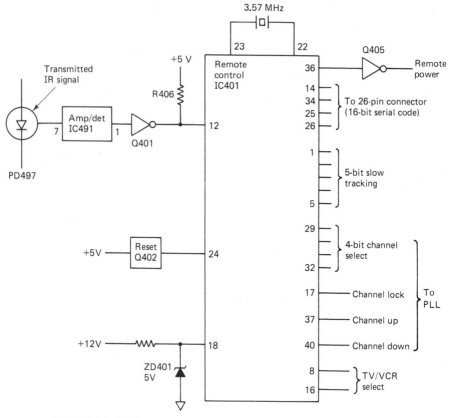

FIGURE 4-6 Remote-control decoder microprocessor circuits

operation of which is discussed in Chapter 14. The outputs at pins 17, 29, 32, 37, and 40 of IC401 are the channel-select commands; they are discussed in Chapter 10. The remaining outputs are discussed in this chapter.

Remote-control decoder troubleshooting/repair. If you do not get proper remote-control operation, substitute a known good remote transmitter. Next, confirm the presence of the signal at pin 12 of IC401. This signal is a 32-bit serial data train that cannot be intelligently monitored by an oscilloscope or probe. That is, there is no easy way to determine which logic code is being transmitted by monitoring the signal. However, it is reasonable to assume that if a signal is present, the code is correct. If the signal is present, check for 5 V (the B+ supply) at pin 18 of IC401. If present, check for a 3.57-MHz clock signal at pin 23 of IC401, and for a reset high at pin 24 of IC401. If these inputs are correct, but there is no remote-control operation (but manual operation of the VCR is correct), suspect a defective IC401.

If you get remote-control operation for some functions, but not all functions, check the corresponding output from IC401. For example, if you get channel-up operation but no channel-down operation (with a known good remote transmitter), check for a channel-down output at pin 40 of IC401. If the output is absent or abnormal, suspect IC401. If the output is present, the problem is likely in the tuning system and should be checked following the procedures of Chapter 10.

4-5 TV/VCR-SELECT CIRCUIT

Figure 4-7 shows the TV/VCR-select circuits for the VCR of Fig. 1-1c. In this VCR the RF antenna-switching circuit is of the automatic type rather than the manual detented switch often found on early-model VCRs. The RF antenna-switching relay and circuits are controlled by remote-control IC401 (pins 8 and 16). The IR remote transmitter, or the front-panel TV/VCR switch, control the RF antenna function. The TV/VCR switch is connected to pin 16 of IC401. When the VCR is operated in the play mode, or when the TV/VCR switch is pressed, a logic high is applied to pin 16 and IC401 "toggles" the output at pin 8. (If pin 8 is high, it goes low, and vice versa. Pin 8 is high during VCR and low during TV mode.)

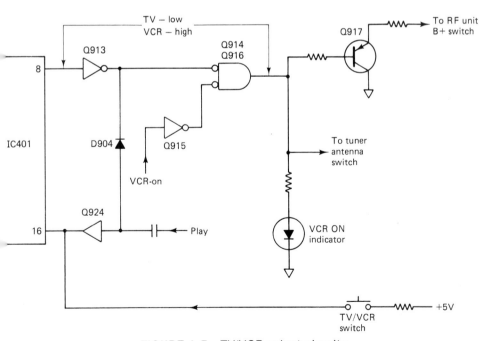

FIGURE 4-7 TV/VCR-select circuits

In the VCR mode, with pin 8 high, the output of Q913 is driven low. This low is applied to the inverting input of Q914/Q916. When the VCR is turned on, a high is applied to Q915, inverted, and then applied to the second inverting input of Q914/Q916. When these two lows are applied, the output of Q914/Q916 goes high. This gating arrangement allows automatic return to the TV mode for the antenna switch when the VCR is turned off. (If there is no VCR-on signal to Q915, the output of Q914/Q916 is low, and the TV is connected to the external antenna or cable.)

The high at the output of Q914/Q916 is applied to Q917 to provide the RF modulator with B+. The high at Q914/Q916 is also applied to the RF antenna switch (connecting the TV to the VCR output) and to the VCR-on LED on the front panel.

TV/VCR-select troubleshooting/repair. There are three general symptoms of TV/VCR-select circuit problems.

If you get no automatic VCR selection during play mode (that is, if you get both manual and remote operation, but not automatically when the VCR is set to play), suspect Q924.

If you get remote control operation of the TV/VCR-select, but no manual operation, check for a high at IC401-16 when the TV/VCR switch is pressed. If you get the high, but no response to manual (front-panel) commands, suspect IC401.

If you get no TV/VCR switching from remote or manual operation, or when the VCR is set to play, check that pin 8 of IC401 toggles between high and low when the corresponding modes are selected. If IC401-8 does not toggle each time the TV/VCR switch is pressed, suspect IC401. If IC401-8 does toggle, then check for defective transistors Q914, Q915, and Q916. If the antenna switch does change between the external antenna and the VCR, but B+ is not applied and removed from the RF modulator, check Q917. If the opposite occurs, check the wiring between the output of Q914/Q916 and the antenna switch.

4-6 SLOW-TRACKING CONTROL GENERATION

Figure 4-8 shows the slow-tracking control generation circuits for the VCR of Fig. 1-1c. The 32-bit slow-tracking signal received by IC401 from the remote transmitter is decoded to a 5-bit signal appearing at pins 1 through 5 of IC401. This 5-bit output is connected through a resistor ladder network and can generate 32 different voltage levels at IC402-8. A 2-kHz sawtooth waveform from IC402 and Q403 is applied to pin 9 of IC402. The combination of the two signals (5-bit and sawtooth) determines the pulse width of the signal at IC402-14. This variable-width square wave (or pulse width modulated, PWM) signal is filtered by R403/C409 to an average dc value. The value appearing across

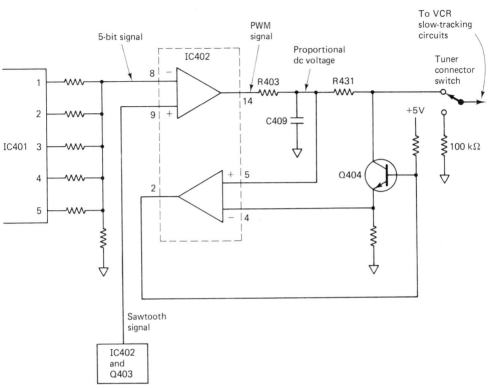

FIGURE 4-8 Slow-tracking control generation circuits

C409 is directly proportional to the pulse width of the square wave signal. The dc voltage across C409 is applied to pin 5 of IC402.

The output of the op-amp at IC402-2 is applied to the base of Q404. The emitter of Q404 is tied to pin 4 of IC402 to provide closed-loop feedback. This arrangement produces a voltage-to-current conversion of the signal appearing across C409. The combination of R431, IC402, and Q404 form a constant-current generator to maintain constant current through R431. By keeping this current constant, the voltage variation across C409 generates various conduction states (or resistance variations) in Q404.

The conduction state of Q404 (the resistance value produced by Q404) is applied through a rear connector to the VCR slow-tracking circuits. (When the VCR is connected to the tuner/timer/charger, an internal sensing switch within the connector is activated to apply the output to the slow-tracking circuits.) The resistance or conduction of Q404 determines the slow-tracking control for the VCR. When the VCR is removed from the tuner/timer/charger, the connector switch is toggled, connecting a 100-kΩ resistor to the slow-tracking circuits. (This develops a nominal slow tracking during portable operation of the VCR.)

Slow-tracking control troubleshooting/repair. If you do not get slow-motion tracking control during slow-motion operation, check the remote control transmitter by substitution. Then check for the 5-bit signal appearing at pins 1 through 5 of IC401. If the signals appear, then check the range of voltage appearing at pin 8 of IC402 between minimum and maximum tracking adjustments. (The voltage should be 0 at minimum and about 1.3 V at maximum). If the voltage is incorrect, suspect a defective IC401. If the voltage range is correct, check the sawtooth waveform at IC402-9. If missing, suspect the sawtooth generator circuits of IC402 and Q403.

If both the voltage range at IC402-8 and sawtooth voltage at IC402-9 are present and correct, check for PWM square wave signals at IC402-14. If missing, suspect IC402. If present, monitor the minimum and maximum voltage appearing at the collector of Q404. (The minimum is 0.3 V, with about 3.4 V maximum.) If the voltage is incorrect, or possibly latched to either a high or low state, suspect IC402 and/or Q404. If the voltage at the collector of Q404 is changing correctly as the remote-control transmitter slow-tracking control is operated, but there is no slow-tracking control evident in the playback picture, suspect the tuner connector switch (or the slow-tracking circuits within the VCR, Chapter 9).

4-7 TUNER/TIMER/CHARGER COMMUNICATIONS TO VCR

As discussed in Chapter 14, the data communications link between the tuner/timer/charger and the VCR is made through a 16-bit serial bus, including appropriate handshaking signals. These signals are passed through the 26-pin connector at the rear of the VCR, as shown in Fig. 4-9.

For any command function of the remote-control transmitter that affects the VCR mode of operation, the corresponding signal is passed from IC401 to the VCR system control microprocessor IC801 (Chapter 14). The communications sequence between IC401 and IC801 is as follows:

1. IC401 pulls the READY-2 line low.
2. IC801 receives the ready signal and pulls the ACK (acknowledge) 1.2 line low.
3. IC401 receives the acknowledge signal, outputs the first of the 16-bit data signals, and returns the READY-2 line high.
4. IC801 receives the data bit and returns ACK 1.2 line to high.
5. The sequence repeats to completely send the 16-bit signal.

During manual operation of the VCR, the requested mode information is also relayed back to IC401.

Tuner/Timer/Charger Communications to VCR 113

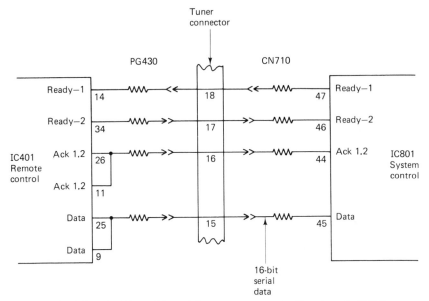

FIGURE 4-9 Tuner/timer/charge communication bus to VCR

Remote control communications troubleshooting/repair. It is not practical to monitor the 16-bit data signals and handshake signals (ready/acknowledge) with a conventional oscilloscope or logic probe. (RCA uses a special serial decoder test fixture to monitor the signals.) However, if the signals are present, it is reasonable to assume that the signals are correct. So if you get good manual operation of the VCR, but no remote control operation, proceed as follows:

First check the inputs to IC401 as described in Sec. 4-4. If the inputs are good, check for the presence of a 16-bit READY-2 signal at pin 34 of IC401 when a remote command is given. If the READY-2 signal does not appear, suspect IC401. If the signal appears, check for the presence of the ACK 1.2 signal at pin 11 or 26 of IC401. If missing, suspect a problem with the system control microprocessor IC801, or a defect in the bus wiring. If there is an ACK 1.2 signal at pins 11 and 26 of IC401, check for the presence of a data signal at pin 9 or 25 of IC401. If data bits appear to be present, but there is still no remote operation, the problem can be in either IC401 or IC801 (with IC801 the most likely suspect). Of course, check all bus wiring before you replace either IC801 or IC401.

5

THE MECHANICAL SECTIONS

This chapter describes troubleshooting and repair for the mechanical sections of typical VCRs. Both top-load and front-load VCRs are covered. Since operation of the mechanical sections (tape transport, cassette load/unload, tape load/unload, and safety devices) are closely related to the system control functions, it is essential that you review the corresponding system control descriptions of Chapter 14.

By studying the mechanical operation found here, you should have no difficulty in understanding the mechanical operations of similar VCRs. This understanding is essential for logical troubleshooting and repair, no matter what VCR is involved. For example, if you know that the reel brakes are actuated in the stop mode (typical for most VCRs), and you see that the reel brakes are not actuated, you have pinpointed a failure symptom. The origin of the trouble may be electronic (no reel-brake-actuating signal is present) or mechanical (the reel brake is jammed), but you have a starting point for troubleshooting. The descriptions given here should also help you interpret the mechanical sections of VCR service literature (which is usually well illustrated, but often poorly described).

5-1 TYPICAL TOP-LOAD MECHANICAL OPERATION

The following paragraphs describe mechanical operation of a typical top-load VCR (which happens to be a Beta unit).

5-1.1 Major elements of the top-load mechanical section

The mechanical section or mechanism of a top-load VCR can be divided into two blocks: *drum block* and *chassis block*. The drum block is composed primarily of the tape travel system (drum with the rotating video heads, erase head, audio/control head, tape guide, capstan, pinch roller, loading ring, etc.). The tape travel or drum block system is adjusted at the time of manufacture and assembled into a complete functioning unit. The chassis block is composed of the supply and take-up reel base, capstan motor, reel motor, solenoids, cassette holder, and related parts.

The drive system for the mechanical section is composed of dc motors: the direct-drive (DD) *head motor* for video head rotation (installed on the drum); the *capstan motor,* which drives the capstan for tape feed; and the *reel motors,* which drive the reels for take-up and rewinding of the cassette tape. Both the head motor and capstan motor are part of servo systems, and are discussed in Chapters 7 and 8, respectively. The reel motors are covered in this chapter. Note that the top-load mechanism described in this section has one reel motor, connected to the reels by belts, whereas the front-load mechanism of Sec. 5-2 has two DD reel motors (one for the supply and one for the take-up reel). In the top-loading mechanism described here, turning of the loading ring during load and unload of the tape is executed by the capstan motor via a *capstan flywheel.* In the front-load mechanism (Sec. 5-2) the loading ring is driven by a *tape-loading motor.* The top-load mechanism also has the following solenoids: the *pinch roller solenoid* for pinch roller attraction, the *brake solenoid* for brake release, and the *roller solenoid* for roller attraction. Mechanism switching for the various operating modes is executed by combined operation of the three solenoids.

5-1.2 Top-load tape run system

Figure 5-1 shows mechanical parts of the top-load tape run system. As shown, the tape run system consists essentially of the drum, the tape guides around the drum, and the tape guides installed on the loading ring. The tape guide (2) of the forward sensor assembly (1) is arranged so that a constant distance is kept between the forward sensor assembly (3) and the tape.

The back-tension lever (4) guides the tape so that the tape runs in the direction toward the drum, and the tension of the brake band (5) is adjusted so that a constant braking force acts on the supply reel base for constant back-tension on the tape. The back-tension is adjusted by the spring (6), and the installation fitting (7) of this spring (6) plays an important role in the stability of the tape run. The perpendicularity of the back-tension lever (4) has much influence on the tape run from the cassette to the drum.

116   The Mechanical Sections

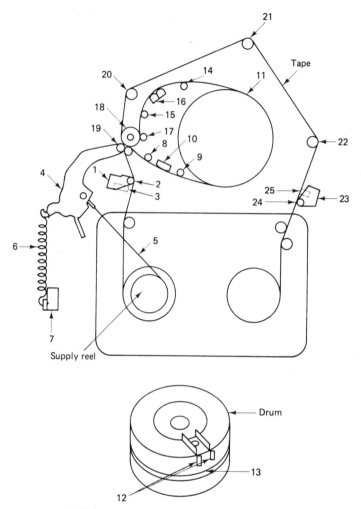

FIGURE 5-1 Top-load tape run system

The erase-head roller (8) absorbs lengthwise vibrations of the tape, caused between the back-tension lever (4) and the erase head (10), and thus removes jitter at the time of playback. The tape input guide (9) controls the upper edge of the tape and guides the tape so that correct tape run is executed. In record, the full-erase head (10) erases all signals from the passing tape.

The tape is wound around the drum (11) at an angle along the groove (13). The tape wraps the drum by 180° + alpha (where alpha is the amount of overlap). The rotating video heads scan the tape and execute recording and playback of the video signals. The tape push lever (12), installed on the top of the drum, uses a spring to push the tape (at the approximate center of the drum) downward. As a result, the tape can run correctly along the groove (13) at the bottom of the drum. The tape outlet guide A(14) regulates the upper edge of the tape, while tape outlet guide B(15) regulates the lower edge of the tape, so that the tape coming out from the drum (11) is a stable run.

The tape inlet guide (9) and tape outlet guide A(14) regulate the angle (and amount of wrap) at which the tape is wound around the drum.

The audio/control head (16) executes recording and playback of the audio signal. Generally, the height and angle of head (16) are adjusted at the factory and need not be changed unless the head is replaced. Head azimuth and tracking position adjustments are performed after assembly of the mechanism. (We describe typical adjustments at the end of this chapter.)

The capstan (17) has considerable influence on tape speed and the stability of tape run. For this reason, the capstan (17) has a precision finish to eliminate out-of-roundness, rectangularity, and surface roughness. The pinch roller (18) is pushed at a right angle against the capstan (17), to provide the tape with a uniform pressure and to ensure stable tape run.

Guide (19), loading guide A(20), and loading guide B(21) are roller-type guides used to reduce the friction load. The loading guide C(22) compensates for the plane angle between the cassette plane and the loading ring plane, and holds the tape so as to guide the tape from the loading guide B(21) to the cassette. The tape guide (24) of the rewind sensor assembly (23) is arranged so that a constant distance is kept between the rewind sensor (25) and the tape.

5-1.3 Top-load operation in the loading mode

Figure 5-2 shows operation of the top-load mechanical parts in the loading mode. When the cassette is inserted into the cassette holder (1) and is pushed down, the synchronization gears (2) and (3) engage with the racks on the top-load cassette holder (1) and are turned in the arrow direction. The damper gear (4), engaged with the synchronization gear (3), also turns in the arrow direction. The damper unit (5) is installed on the same shaft as the synchronization gear (2). When the synchronization gear (2) rotates, the damper unit shifts in the arrow direction, is disengaged by stopper (6), the rotation of damper gear (4) becomes unbraked, and the cassette holder (1) can be pushed in quickly. The cassette holder lock cam (7) is installed on the end of the synchronization gears (2) and (3), and rotates in coincidence with the rotation of the synchronization gears. The ejection spring (9) is pulsed via the attached wire (8). The cassette holder (1) is locked by catching of the cassette holder lock lever (10) at the cam part.

When the cassette holder lock lever (10) is shifted, the cassette holder pulley lock lever (11), held at part A, is shifted in the arrow direction, and this shifts the eject link (12) held by part B. The cam (13) is installed on the eject link (12), and the eject switches (14) and (15) turn on by movement of the eject link (12). The eject link (12) is connected to the loading lever (16) at

118 The Mechanical Sections

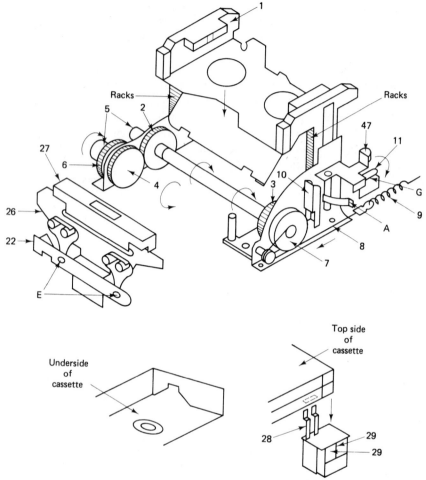

FIGURE 5-2 Top-load loading mode operation

part C and moves the lever in the arrow direction. The loading lever (16) cancels the control of the loading gear assembly (18), pulled by spring (17). The loading gear assembly (18) is turned in the arrow direction, and the loading roller (19) is pushed against the flywheel (20).

The eject link (12) also moves the ejection operation link (22) by the eject lever (21) in the arrow direction. At part D, the eject operation link (22) shifts the eject brake lever (23) in the arrow direction, and in order to release the brake from the supply reel base (24), the eject operation lever (25) is turned at part E to return the eject operation slide (26) and the eject lever (27) in the arrow direction.

When the cassette holder (1) is locked, the cassette pushes the cassette

Typical Top-Load Mechanical Operation 119

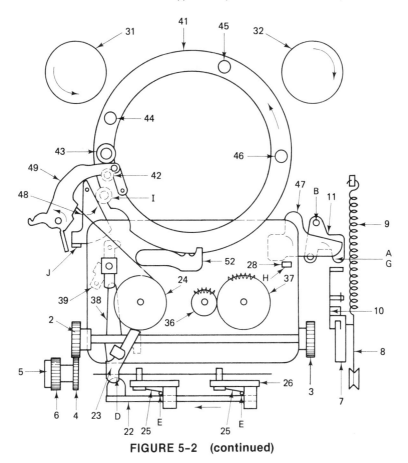

FIGURE 5-2 (continued)

switch actuator (28) and the cassette switch (29) turns on. Also, the eject switches (14) and (15) turn on, and head motor (30), capstan motor (31), reel motor (32), and roller solenoid (33) are engaged by these switches. Note that when the cassette holder (1) is locked without insertion of a cassette, the cassette switch (29) does not go on, so that only eject operation becomes possible.

The roller solenoid (33) attracts in the arrow direction and shifts the forward roller lever (35) via link (34). The forward gear (36) is removed from the take-up reel base (37), and the drive of the reel motor (32) is not transmitted to the take-up reel base (37).

The forward roller lever (35) turns the supply reel base brake lever (38) at part F in the arrow direction and releases the brake of the supply reel base (24). However, the soft brake lever (39) is applied at the supply reel base (24) for slight braking to prevent tape slack at the time of loading.

The rotation force of the capstan motor (31) is transmitted via the belt

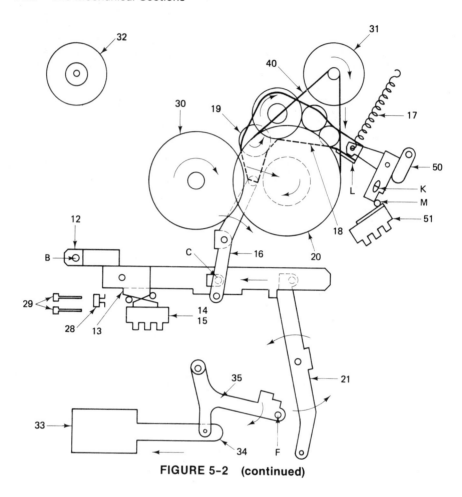

FIGURE 5-2 (continued)

(40) to the flywheel (20) and further from the loading roller (19) to the loading gear assembly (18). The loading ring (41) (engaged with the loading gear) is turned in the arrow direction, and guide (42), pin roller (43), and loading guides (44), (45), and (46) pull the tape from the cassette to execute loading.

The cam part of the loading ring (41) turns the unloading end-sensing lever (47), the cassette holder lock lever (10) is pushed at part G, the cassette holder (1) is pushed down further from the prelock position, and the cassette is locked. The unloading end-sensing lever is controlled by the cassette switch actuator (28) at part H, and when the cassette holder (1) is locked without insertion of a cassette, actuator (28) does not move to release the control, so that the unloading end-sensing lever (47) locks the loading ring (41) at the cam part to prevent unnecessary turning.

The cam I of the loading ring (41) shifts the loading end-sensing lever (48), which controls the spring-pulled back-tension lever (49) at part J. The shifting by rotation of the loading ring (41) loosens the control release of the back-tension lever (49) for shifting in the arrow direction to pull the tape from the cassette.

When the loading end-sensing lever (48) engages in the recessed part of cam I of the loading ring (41), the back-tension lever (49) is released from the control by part J. The loading end-sensing lever (48) turns the loading gear cancellation lever (50), connected at part K, in the arrow direction. The loading gear cancellation lever (50) pushes the loading gear assembly (18) at part L to cancel the pressure of the loading roller (19) and flywheel (20), and to stop rotation of the loading ring (41). At the same time, the loading end switch (51) is actuated to ON by part M, and the energizing of motors and solenoids is stopped. The loading ring (41) is stopped by stopper (52). Supply-reel base-brake operation and forward-gear pushing operation are executed by attraction of the roller solenoid (33). The STOP mode is reached with completion of loading.

5-1.4 Top-load operation in the PLAY mode

Figure 5-3 shows operation of the top-load mechanical parts in the play mode. In the play mode, the head motor (1), capstan motor (2), and reel motor (3) are running, and the pinch roller solenoid (4) is attracted.

With the pinch roller solenoid (4) attracted, the pinch roller pressure lever (8) and the pressure operation level (9) are shifted in the arrow direction via the links (5), (6), and (7). The pinch lever (10) is pushed by part A of the pinch roller pressure lever (8), and the pinch roller (11) is pushed against the capstan (12).

The pressure spring (13) is applied to the pressure operation lever (9), and pressure is applied by the force of this spring. At part B, the pinch roller pressure lever (8) turns the brake release lever (14), the coupled soft brake lever (15) is turned in the arrow direction, and soft brake release is executed. At part C, the soft brake lever (15) shifts the supply reel brake lever (16) in the arrow direction and executes brake release for the supply reel base (17).

The rotation of the capstan motor (2) is transmitted by the belt (18) to the flywheel, and the tape is transported with constant speed by the capstan (12) and the pinch roller (11). The capstan motor (2) is controlled so that the tape runs with a speed of 20 cm/s in Beta II, and at a speed of 1.33 cm/s in Beta III mode.

The rotation of the reel motor (3) is transmitted to the forward pulley (23) via belt (20), intermediate pulley (21), and belt (22). The forward gear (24), installed as one body with forward pulley (23), turns the take-up reel base (25) in the arrow direction, and the tape paid out by capstan (12) and pinch roller (11) is taken up. The intermediate pulley (21) has a built-in slip

122   The Mechanical Sections

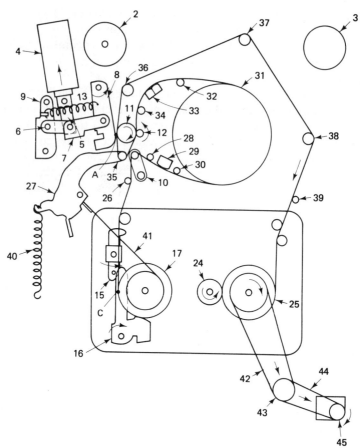

FIGURE 5-3 Top-load PLAY mode operation

mechanism, which controls the rotation force of the reel motor (3) so that the tape is taken up with a tension force that does not damage the tape.

The tape is pulled from the cassette and passes the forward sensor guide (26), back-tension lever (27), erase head roller (28), erase head (29), inlet guide (30), drum (31), outlet guide (32), audio control head (33), and outlet guide (34). The tape is then transported by pinch roller (11) and capstan (12), passing guide (35), loading guides (36), (37), (38), and rewind sensor guide (39) to be taken up into the cassette.

The back-tension lever (27) pulls the brake band (41) with the spring (4) to brake the supply reel base (17) so that a constant back tension acts on the tape. The rotation of the take-up reel base (25) is transmitted via belt (42), pulley (43), and belt (44) to the tape counter (45) for tape position display.

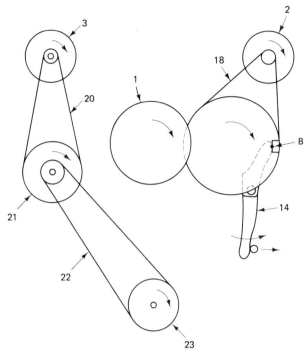

FIGURE 5-3 (continued)

5-2 TYPICAL FRONT-LOAD MECHANICAL OPERATION

The following paragraphs describe mechanical operation of a typical front-load VCR (which happens to be the VHS unit shown in Fig. 1-1a).

5-2.1 Front-load tape transport and status sensors

Figure 5-4 shows the major components of a typical front-load tape transport. The tape transport deck performs all the mechanical functions related to tape handling and movement. When the front-load cassette is locked into position, the tape transport withdraws the tape from the cassette and routes the tape through a system of precision alignment posts and guides, precisely aligning the tape for application to the erase, video, audio, and control track heads. The tape transport also supplies the drive necessary to pull the tape from the cassette, through the transport assembly, and return the tape back into the cassette. Throughout this process, electronic sensors continually monitor the condition of the tape and freedom of movement through the transport. All of these functions are under control of microprocessors in the system con-

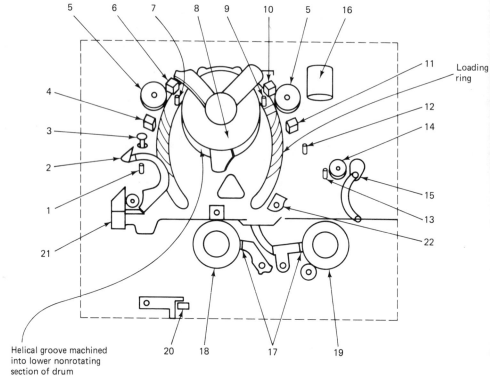

FIGURE 5-4 Major components of a typical front-load tape transport

trol circuits (Chapter 14). The following is a brief description of the operating sequence for the tape transport.

All active transport elements are positioned by a *loading ring*. The use of a loading ring for VHS is one feature that distinguishes most modern VCRs from early-model VHS units. The loading ring is responsible for the initial withdrawal of the tape from the cassette, as well as directing the movable guides to their final positions.

Exiting the cassette, the tape is initially positioned by a fixed guide post (1), contacts the supply tension arm (2), and is further positioned by a larger-diameter supply guide post (3) with upper and lower positioning flanges. The tape then moves past the full-erase head (4) and pressure idler (5) which supplies the tension necessary to maintain tape contact with the input guide roller (6). The final tape position is refined by the input slant-guide post (7), to provide the proper approach angle to the drum assembly (8). The lower edge of the tape is guided around a portion of the drum by a precision helical groove machined into the lower, nonrotating section of the drum.

The exit slant-guide post (9) positions the tape for departure from the drum, and the tape is applied to the take-up guide roller (10), past the audio/control head (11) and fixed take-up guide post (12), to the capstan drive shaft (13), where the tape is held against the shaft by the capstan pinch roller (14). The capstan drive shaft provides the momentum necessary to move the tape through the transport at a consistent predetermined speed.

Upon leaving the transport deck, the tape is prepared for reentry into the cassette by the take-up tension arm (15). The loading motor (16) supplies the power to drive the loading ring, located underneath the transport place, via a belt-and-pulley coupling.

Additional elements of the transport include the mechanical reel brakes (17), activated when the VCR is in the stop mode, and the electronically controlled supply reel (18) and take-up reel (19) with their associated drive motors (Sec. 5-2.4 and 5-2.5).

Status switches inform the mechanical control microprocessor in the system control (Chapter 14) of the progress of certain cassette and tape-handling procedures. For example, during tape loading or unloading processes, the P/R (play/record) switch S572 informs system control that tape loading or unloading has been completed, terminating drive to the tape loading motor (which drives the loading ring). The F/R (forward/reverse) switch S573 detects the position of the reel brakes, either on or off. Both the P/R and F/R switches are mechanically activated by, and located adjacent to, the loading ring, underneath the transport base plate.

The record inhibit switch (20) prevents recording on a cassette where the erasure-prevention tab has been removed. As discussed in Sec. 1-9, when the tab is removed, the system control ignores record commands (from the front panel or remote control) and the original recording is preserved.

The supply tension sensor (21) detects tension on the tape just as the tape leaves the supply reel. As discussed in Sec. 5-2.11, sensor (21) controls the supply reel motor voltage, thus varying the back tension as required for uniform tape tension throughout the transport path.

Sensors, located on the transport deck, provide status information inputs to the mechanical control microprocessors in the system control. Sensor inputs to system control have priority over all other inputs. For example, the cassette lamp (22) works in conjunction with the *start sensor* (Sec. 5-2.18) and *end sensor* (Sec. 5-2.17) to detect the transparent leader at the beginning and end of the cassette, placing the VCR in the stop mode. (In certain cases, this is followed by the rewind mode.) If a cassette is not inserted in the VCR, both start and end sensors are activated, inhibiting mode selection and maintaining the VCR in the stop mode. As discussed in Sec. 5-2.21, the *tape slack sensor* detects any tape spillover, placing the VCR in stop mode. As discussed in Sec. 5-2.20, the *dew sensor* responds to a buildup of condensation on the drum by automatically engaging the stop mode.

5-2.2 Front-load cassette mechanism

Figure 5-5 shows the major components of a typical front-load cassette mechanism. Front loading allows the VCR to be stored and operated in a much smaller, confined space. The partial insertion of a cassette into the cassette housing or assembly automatically activates the front-loading (FL) mechanism, drawing the cassette into the VCR and gently lowering the cassette into position on the transport deck. The necessary driving power is derived from the front-loading motor, worm gear, and drive gear assembly shown in Fig. 5-5. The drive gear may be driven in either of two ways during a cassette

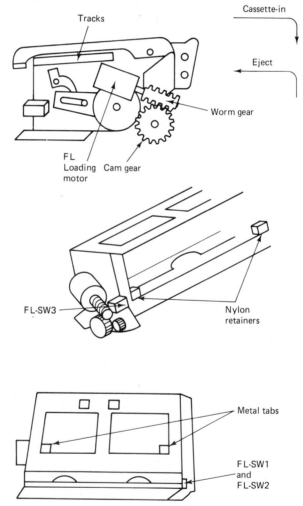

FIGURE 5-5 Major components of a typical front-load cassette mechanism

loading sequence: *manually,* as the cassette is partially inserted by hand, followed by a *powered* drive sequence from the FL motor. Switches FL-SW1, FL-SW2, and FL-SW3 shown in Fig. 5-5 are responsible for initiation and termination of cassette loading/unloading processes.

Figure 5-6 shows the circuits involved in the front-load cassette load/unload/eject operation. Figure 5-7 is the corresponding timing chart. The following is a brief description of the operating sequence for a typical front-load cassette mechanism.

The drive gear simultaneously rotates both the cam gear shown in Fig. 5-5 and a similar cam gear on the right side of the FL mechanism housing. Rotation of the cam gears serves a dual purpose: (1) it moves the cassette along tracks, or channels, machined into the exterior of the FL mechanism (the tracks predetermine the horizontal and vertical movement of the cassette); (2) it activates, at prescribed points during the cassette loading or unloading process, three status switches (located adjacent to the cam gears) responsible for enabling the FL motor and terminating both the load and unload cycles.

Each status switch is responsible for initiation of a particular segment of the loading or unloading procedure:

Start switch FL-SW1 begins the powered segment of the loading process.

Load stop switch FL-SW2 terminates the load.

Unload stop switch FL-SW3 terminates unload.

The FL mechanism is not operational unless the VCR power switch is on. As a result, certain operating characteristics of the FL mechanism must be noted.

The VCR power switch must be on to load, unload, or eject a cassette.

During cassette load or unload, all VCR operational modes are inhibited, with the exception of eject.

If eject is engaged during an active cassette-loading process, the FL mechanism simply reverses and immediately unloads the cassette.

Eject is inhibited during the record mode.

If the VCR power switch is turned off during a loading or eject cycle, the power remains on until the cycle is completed.

If the FL mechanism jams or stops for any reason, drive to the FL motor is removed after approximately five seconds. The cassette may then be ejected by pressing the eject button.

Cassette load operation. As shown in Fig. 5-6, energy for the powered rotation of the drive components of the FL mechanism is obtained from the FL motor, driven by the FL motor drive IC5A3. In turn, IC5A3 receives control logic from a mechanical microprocessor IC5A1 (part of system control,

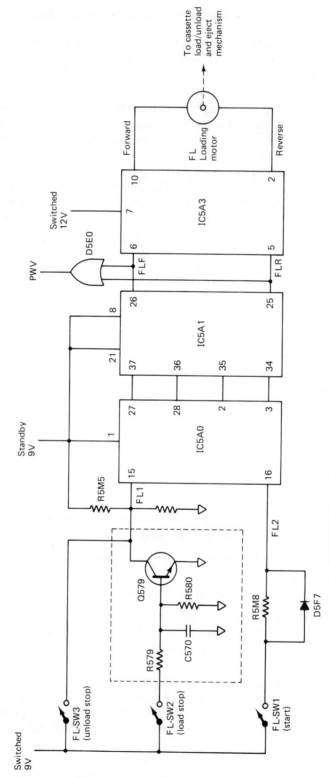

FIGURE 5-6 Front-load cassette load/unload/eject circuits

128

Typical Front-Load Mechanical Operation   129

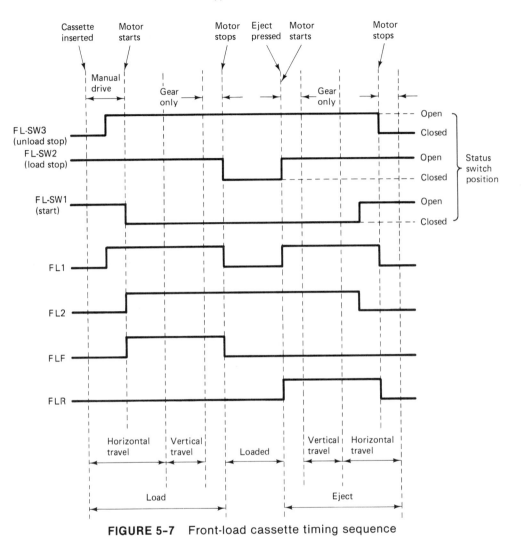

FIGURE 5-7 Front-load cassette timing sequence

Chapter 14). A high on the FLF (front-loading forward) line, from pin 26 of IC5A1, places a high on pin 10 of IC5A3, driving the FL motor in a forward direction. Conversely, a high on the FLR (front-loading reverse) line, from pin 25 of IC5A1, places a high on pin 2 of IC5A3, driving the motor in a reverse direction.

IC5A1 interfaces with, and receives logic from, expander IC5A0. The status at any point in the cassette load or unload process is detected, and the input to IC5A0 is determined by the open or closed condition of the start FL-SW1, load-stop FL-SW2, and unload-stop FL-SW3 status switches.

When the FL system is at rest, in a cassette-loaded condition, FL-SW1 and FL-SW2 are open and FL-SW3 is closed. Q579 is not conducting, and the FL1,FL2 expander inputs (pins 15 and 16) are low. The cassette housing is latched in position by two spring-loaded nylon retainers situated one on each side of the lower cassette housing plate, as shown in Fig. 5-5.

The insertion of a cassette into the housing engages two metal tabs at the rear of the housing (Fig. 5-5), releasing the nylon retainers to allow movement of the cassette housing. Continued insertion rotates the right cam gear, closing the start switch FL-SW1 and the left cam gear, opening the unload-stop switch FL-SW3, as shown by the timing sequence in Fig. 5-7. Since Fl-SW2 remains open at this point, Q579 remains off. However, the switching creates a low-to-high transition at the FL1 expander input (pin 15) from the standby 9-V supply and R5M5, and at the FL2 input (pin 16) from the switched 9-V supply, FL-SW1, and R5M8/D5F7. Corresponding outputs from IC5A0 are decoded by IC5A1, producing a high on the FLF line applied to pin 6 of IC5A3. Pin 10 of IC5A3 then goes high, driving the FL motor in a forward direction.

The FL motor then rotates the various gears, moving the cassette into the VCR along the horizontal tracks of the FL housing, then along the vertical tracks (as indicated by the cassette-in arrows of Fig. 5-5), gently lowering the cassette onto the hubs of the take-up and supply reels. At this point the FL motor continues to operate, rotating the gears an additional 10°, locking the cassette into place. When the cassette is in place, FL-SW2 closes, turning on Q579 and creating a high-to-low transition at the FL1 expander input. Through action of IC5A0, IC5A1, and IC5A3, this transition causes the FLF line to go low and terminates FL motor drive.

Cassette unload/eject operation. An eject command can be completed during any operating mode except record. If eject is selected during a tape-loaded mode (Sec. 5-2.3), IC5A3 places the VCR in stop and, when the tape is fully unloaded, implements eject by generating a high on the FLR line to drive the FL motor in reverse.

Initially (as the FL motor starts to reverse), only the drive gear rotates, unlocking the cassette from the loaded position. Once the cassette is unlocked, the cam gears begin to rotate, moving the cassette vertically and then horizontally toward the front of the VCR (as indicated by the eject arrows in Fig. 5-5). During eject, any low-to-high logic transition at the FL1 and FL2 inputs to IC5A0 are ignored. Only a high-to-low transition at either FL1 or FL2 is recognized.

As shown in the timing chart of Fig. 5-7, FL-SW2 opens simultaneously with motor start. This turns off Q579 and allows FL1 to go high. However, the low-to-high transition is ignored by IC5A0. Prior to completion of the eject cycle, FL-SW1 also opens, driving the FL2 input low. IC5A0 is then prepared to accept the final command. As the cam gear continues to rotate, FL-SW3 closes, creating the required high-to-low transition at FL1. Through action of IC5A0, IC5A1, and IC5A3, this transition causes the FLR line to go low and terminates FL motor drive.

Unload power hold-on. If the VCR power switch is turned off during a cassette load or eject cycle, switched power (Sec. 3-1.1) is maintained until the cycle has been completed by holding the PWV (power command) line high. During the loading process, the FLF line (IC5A1-26) is high, holding the PWV line high via the OR gate of D5E0. During eject, the PWV line is held high by the FLR line (IC5A1-25) via D5E0. In either case, the switched B+ supplies remain on until the PWV line goes low at the end of the cassette load or eject.

5-2.3 Front-load tape loading/unloading mechanism

As discussed in Sec. 5-2.1, Fig. 5-4 shows the major components found in the mechanical section of a typical front-load VCR. Figure 5-8 shows both the circuits and timing involved for operation of the tape load/unload mechanism. Note that the tape load/unload function uses the same control system

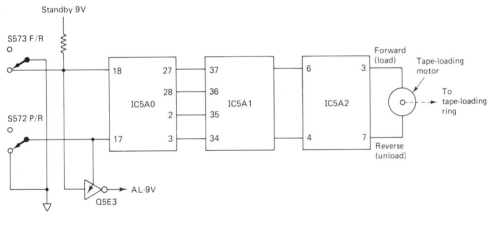

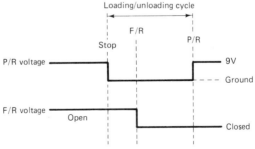

FIGURE 5-8 Tape load/unload circuits and timing

ICs as the cassette load/unload/eject functions, but there is no mechanical connection to the eject function. This is in contrast to the top-load VCRs (Sec. 5-1) where load/unload/eject operations are directly related (mechanically coupled and driven by the same motor). This separation of cassette eject from tape load/unload simplifies operation of the tape load/unload functions of an FL VCR.

The circuits of an FL VCR control power to the motor, which drives the loading ring. In turn, the loading ring operates the mechanical components responsible for tape withdrawal from the cassette, positioning of the tape in the transport, and tape reinsertion back into the cassette. The loading ring is driven by a dc motor powered from the loading motor drive IC5A2. In turn, IC5A2 is controlled by IC5A0 and IC5A1, the same ICs used by the cassette load/unload/eject mechanism.

Movement of the loading ring, and its position throughout a tape loading or unloading cycle, is monitored by two status switches located adjacent to and activated by the loading ring. The P/R (play/record) switch S572 detects the loaded or unloaded condition of the tape and terminates loading motor drive at the proper time. The F/R (fast forward/rewind) switch S573 detects the application of the reel brakes (and is used primarily during the fast-forward and rewind modes). Note that IC5A1 responds differently to status switch logic, depending on the operational mode.

Tape loading. Tape loading occurs only during play or record modes. An initial high is generated on the load line (IC5A1-28), establishing a high at IC5A2-3, and forward rotation of the motor/loading ring. This high remains indefinitely until a low-to-high transition occurs at the P/R input (IC5A0-17). The load line then drops to low and motor drive is stopped. As shown by the timing chart of Fig. 5-8, when loading starts, the loading ring closes S572 to ground, producing a low at the P/R input. As loading progresses, the low is maintained. At the completion of tape load, the loading ring closes the P/R switch to the standby 9-V terminal, resulting in the low-to-high transition at the P/R input necessary to stop the motor/loading ring.

The F/R switch S573 is used primarily during FF (fast-forward) and REW (rewind) modes. During a play/record loading or unloading cycle, S573 has no effect on IC5A0 and IC5A1. However, S573 has direct effect on transistor Q5E3, which controls the AL-9V (after loading 9V) signal necessary to start play or record when the loading is complete.

Again referring to the timing chart of Fig. 5-8, in the fully unloaded condition (stop mode), both S572 and S573 are closed and connected to a standby 9-V supply. The base of Q5E3 is high from the F/R line, and the emitter is high from the P/R line. Q5E3 is cut off and the AL-9V line at the collector of Q5E3 is zero.

During the loading cycle, when the reel brakes are withdrawn, the loading ring closes S573 to ground, driving the F/R line and the base of Q5E3 low.

Since the P/R line is also low at this point, holding the emitter of Q5E3 low, Q5E3 still does not conduct. At the completion of loading, the P/R line goes high (S572 moves to standby 9V) while the F/R line remains low. Q5E3 is then turned on, generating an AL-9V at the collector output.

Tape unloading. When in play or record, the selection of any mode active in the tape-unloaded condition—such as FF,REW, or stop—generates a high on the unload line (IC5A2-4). This drives the motor/loading ring in the reverse direction. The P/R line drops to a low, turning Q5E3 off, and eliminating the AL-9V signal. At the completion of unloading, S572 produces the low-to-high transition necessary to terminate motor/loading ring drive.

Tape fast-forward/rewind. When FF or REW is selected, reel brakes must be retracted to allow the reel movement necessary for the execution of either mode. Since the position of the reel brakes is determined by movement of the loading ring, drive to the loading motor is supplied for a short time, then terminated at the point of reel brake release.

The selection of FF or REW from a tape-loaded mode first initiates an unloading cycle (as described for tape unloading). This returns the loading ring to the stop position. IC5A1 is programmed such that once the loading ring is at stop, a normal loading cycle is initiated, but with one exception. The status logic at the P/R input is inhibited, while the F/R status logic is recognized.

A high-to-low transition at the F/R input terminates loading motor drive, ending the loading cycle. From the stop position, the loading ring closes S573 to the ground terminal at the point of brake release, creating a high-to-low transition at the F/R input, terminating loading motor drive coincident with the release of the reel brakes. FF or REW then proceeds. At the end of either FF or REW, an unloading cycle is executed, again returning the loading ring to the stop position.

Tape PWV hold-on. If the VCR power switch is accidentally actuated during a cassette load/unload cycle, or during a tape unloading cycle, the switched power is removed and operations cannot be completed. The PWV hold-on circuit shown in Fig. 5-9 maintains switched power until cassette or tape load/unload operations can be completed.

The PWV line carries the power-on command. When PWV is high, the VCR is on, and vice versa. The PWV hold-on circuit of Fig. 5-9 holds the PWV line high until loading and unloading operations are completed, even though the VCR may have been turned off during the procedure. When the VCR is turned on, the PWV line is driven high by D5A1, by a high on the T-PWV line from the timer processor IC8A1 (part of system control). When the VCR is turned off, the T-PWV line goes low, removing the high on the PWV line.

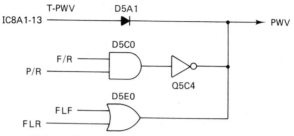

FIGURE 5-9 Tape PWV hold-on circuit

The *front-loading drive* commands, FLF and FLR, are coupled to the PWV line via D5E0. During forward motor rotation (cassette load), the FLF line is high. In eject or cassette unload, the FLR line is high. In either case, the respective high is directed to the PWV line via D5E0, holding the VCR on and maintaining switched power until the cycle is completed and the FLF or FLR lines once again drop to a low.

The two status switches, F/R and P/R, used in the *tape loading mechanism* are tied to the PWV line via the AND gate portion of D5C0 and switching transistor Q5C4. During a tape loading or unloading process, both the F/R and P/R lines are low, turning on Q5C4, and applying standby 9-V power to the PWV line, holding the VCR on. When unloading is complete, both the F/R and P/R lines go high, Q5C4 and the PWV line go low, and switched power is removed.

5-2.4 Reel motor drive systems

In addition to driving the reel motors at a speed consistent with the operating mode, the reel motor drive systems also:

1. Monitor and correct tape back tension.
2. Prevent the reel from going in the wrong direction, contrary to that of the capstan motor (Chapter 8).
3. Provide electromagnetic reel motor braking under certain conditions.
4. Provide protection under a wide variety of possible tape-damaging situations.
5. Help to assure the smooth, unimpeded transition of tape from one reel to the other.

The reel motor systems, both take-up (TU) and supply (SU), accept logic signals from IC5A1, the same IC used by the cassette and tape load/unload systems. Keep in mind that while each reel is capable of rotating in both forward and reverse, each motor is confined to a single direction. The TU motor always rotates clockwise, while the SU motor always rotates counterclockwise.

The TU reel (Fig. 5-4, 19) accepts and winds tape moving through the transport and is driven by the TU motor. So the TU motor is energized only during modes requiring the *forward* movement of tape through the transport (such as play, record, forward search, and fast forward or FF). The TU motor drive is *disconnected during reverse* tape movement modes.

The SU reel (Fig. 5-4, 18), driven by the SU motor, rewinds the tape moving in a *reverse* direction through the transport. So the SU motor is *fully energized* in rewind and reverse search modes. However, *some drive* to the SU motor is maintained in *all forward* tape movement modes (except FF) as well. This partial energizing of the SU motor is necessary to maintain tape back tension at a specified value. This function is provided by the tension sensor circuit (Sec. 5-2.11), an integral part of the SU motor drive system.

Obviously, the drive amplitude for each reel motor must be variable, since the motor speed must be adjusted to match the operating mode. For example, the TU motor must rotate at a much higher speed in FF than in play. The amplitude of the drive voltages is set by the reel motor drive systems, which we will discuss next.

5-2.5 Take-up motor drive

Figure 5-10 shows the circuits for the take-up motor drive system. The TU motor, in series with the emitter ground of Q5E0, is driven by Q599 and Q5E0, connected as a Darlington pair. Q5D8 provides the necessary drive voltage to Q599/Q5E0, thus controlling speed of the TU motor. In addition to controlling motor speed, Q5D8 also provides a switching function to stop the TU motor in reverse operating modes (as discussed in Sec. 5-2.7).

The drive voltage table of Fig. 5-10 shows the various voltages applied to the TU motor in the corresponding operating modes. Higher voltages produce higher speeds, and vice versa. The voltages are set by resistances connected to the emitter of Q5D8. The resistances are selected by command signals from the system control microprocessor IC5A1. As an example, in playback/record (P/R), IC5A1 produces a high at pin 29. This high is divided by R5K1 and R5K5 to a value that produces 1.4 V at the TU motor. In FF mode, which requires a much faster TU motor speed, IC5A1 produces an FF command at pin 31. Although identical to the P/R command, the FF command is applied directly to the emitter of Q5D8, without division by a resistor network, and results in a 5.5-V signal being applied to the TU motor.

In still/pause mode, when tape movement is halted, drive to the TU motor is reduced to a level below that necessary to sustain motor rotation, yet sufficient to maintain a small amount of tape tension. This is done by Q5D5, which connects R5K1 in parallel with R5K2 in response to a still/pause command at IC5A1-16. This produces a 0.5-V drive signal to the TU motor, stopping the motor but keeping some tension on the tape.

136 The Mechanical Sections

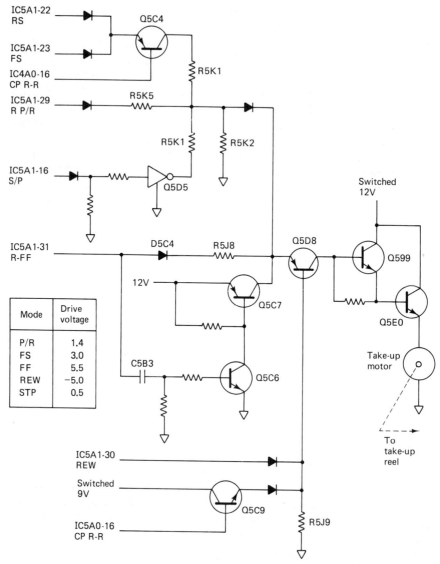

FIGURE 5-10 Take-up reel drive circuits

5-2.6 Forward search (FS) switch

The forward-search switching transistor Q5C4 shown in Fig. 5-10 is used in both the TU and SU motor drive systems. So FS (forward search) and RS (reverse search) outputs are taken from IC5A1. In the TU motor drive system, Q5C4 provides drive to the TU motor during forward search and removes drive from the TU motor during reverse search. Q5C4 functions both as an amplifier and a switch. The logic signal applied to Q5C4 is taken from speed search IC4A0-16 on the CP R-R (capstan reel-reverse) line. The CP R-R logic tracks the rotational *direction* of the capstan motor. The CP R-R logic is low when the capstan motor is in forward, and vice versa. Logic changes on the CP R-R line occur instantly and simultaneously with a change in the rotational direction of the capstan motor.

In the FS mode, the capstan motor rotates in the forward direction, and the CP R-R line is low. This turns on Q5C4. FS logic from IC5A1 appears at the collector of Q5C4 and is applied to Q5D8 through R5K0. This applies approximately 3 V to the TU motor. During RS, the capstan motor rotates in reverse, and the CP R-R line is high, turning off Q5C4. This removes drive voltage to the TU motor, no matter what logic appears on the RS and FS lines.

Transistor Q5C4 also provides protection for the tape in the event of rapid changes from the RS mode to the FS mode. Due to the inertia of the capstan flywheel, the capstan motor does not stop instantly before reversing direction. If the VCR is changed suddenly from RS to FS, the TU motor drive could be applied immediately, causing the TU motor to move the tape forward while the capstan is still in reverse. This could stretch or break the tape.

This problem is avoided by using CP R-R logic as a switching signal for Q5C4. Although the FS logic appears instantly at the emitter of Q5C4, the CP R-R line is high if the capstan is rotating in reverse. This high keeps Q5C4 off temporarily. As soon as the capstan motor begins to rotate in the forward direction, the CP R-R line goes low, Q5C4 conducts, and normal FS TU motor drive is established.

5-2.7 Reverse search (RS)

Transistor Q5D8 also provides a switching function to assure that TU motor drive is removed during reverse tape movement modes. Switching logic for Q5D8 is derived from the REW line, IC5A1-30, and the CP R-R line, IC4A0, through Q5C9. Transistor Q5C9, like Q5C4, is also used in both the TU and SU motor drive systems.

During rewind, TU motor drive is terminated by a high on the REW line, disabling Q5D8. During RS, or when the capstan motor brakes from a forward direction, the CP R-R line goes high, turning Q5C9 on and Q5D8 off to remove drive from the TU motor. *Both* the REW and CP R-R lines must be low to turn on Q5D8 and permit normal TU motor drive.

5-2.8 Rewind brake

Transistors Q5C6 and Q5C7 provide braking action during rewind. If the TU reel continues to rotate at the end of rewind, the tape could pile up. Since the TU motor is being pulled in reverse during rewind, braking is done by momentarily providing forward drive to the TU motor. At the termination of rewind, IC5A1 puts a high on the R-FF (reel-fast forward) line, supplying emitter drive to Q5D8 through D5C4 and R5J8. The R-FF high is also applied across C5B3 as a momentary positive-going pulse to Q5C6. This turns on Q5C6 and Q5C7, applying 12 V to Q5D8, and provides the additional TU drive necessary for rapid braking action.

5-2.9 Rewind electrical brake release

Termination of the TU braking must be timed to coincide when motor rotation ceases to prevent overshoot (continued forward motor rotation). The rewind electrical brake release circuit, shown in Fig. 5-11, informs IC5A1 (at pin 2) that tape movement has ceased, releasing the high on the R-FF line and terminating the braking action.

So long as the supply reel rotates, even after rewind drive is removed, tape movement continues and a positive voltage is generated (inductively) by the SU motor. This positive voltage turns on Q5C2, maintaining a low on the BES (brake end supply) line. When the SU motor stops, signaling the end of tape movement, the positive voltage is no longer generated, Q5C2 is disabled, and the BES line goes high, informing IC5A1-2 that tape movement has stopped. IC5A1 responds by removing the high on the R-FF line, thus removing the TU motor braking.

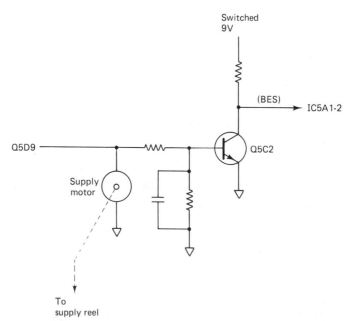

FIGURE 5-11 Rewind electrical brake-release circuit

5-2.10 Supply motor drive

Figure 5-12 shows the circuit for the supply motor drive system. Circuit operation for the SU motor is quite similar to that of the TU motor (Sec. 5-2.5). Transistors Q5D3 and Q571 comprise the Darlington pair drive to the SU motor. Q5D2 serves both as a switching transistor and a dc amplifier, removing drive completely from Q5D3/Q571 (when turned off) or applying drive developed by IC5A1 (when turned on). Note that the voltage ranges applied to the SU motor are similar to those applied to the TU motor (Fig. 5-10), but are applied during different operating modes. Also note that, as shown in Fig. 5-12, the SU motor circuits also include a *tension sensor* and *power failure protect* functions.

5-2.11 Tension sensor

The amount of drive available to the SU motor during the FS, RS, and P/R modes is determined by the tension sensor (21, Fig. 5-4), an assembly consisting of an LED and a phototransistor. The tension sensor is responsible for maintaining tape back tension by altering the amount of SU motor drive in accordance with the amount of tape tension (detected by the sensor arm assembly as the tape exits the cassette). The arm assembly moves freely between the LED and phototransistor, and determines the amount of conduction in the phototransistor. Excessive tension decreases SU motor drive and vice versa.

The tension sensor LED is turned on by Q5E1 in response to a high (from Q5D4/D5D3 during FS, and from D5D5 during P/R). When the LED is turned on, the phototransistor conducts (at a level determined by the position of the tension sensor arm) and supplies a variable bias to Q5E1. The collector output from Q5E1 supplies a variable drive to the SU motor through Q5D2 and Q5D3/Q571. The SU motor drive and tape back tension, are adjustable (by means of VR5A0).

5-2.12 Power failure protect

Transistor Q5C5 provides a power failure protect function to the SU motor drive system. If such a power loss occurs during FF mode, the gear-driven TU reel stops immediately. However, the SU reel continues to rotate (due to inertia) and can result in slack tape. The power failure protect circuit supplies a small amount of drive to the SU motor as a braking force to prevent slack tape.

Transistor Q5C5 is normally turned off due to a high from the standby 9-V supply. Capacitor C4B4 charges to the level of the standby 9-V supply through D5F4. A power loss drives the standby 9 V to zero, placing a low on the base of Q5C5. The charge across C4B4 maintains a high at the emitter of Q5C5, until C4B4 discharges. Q5C5 then conducts momentarily, providing

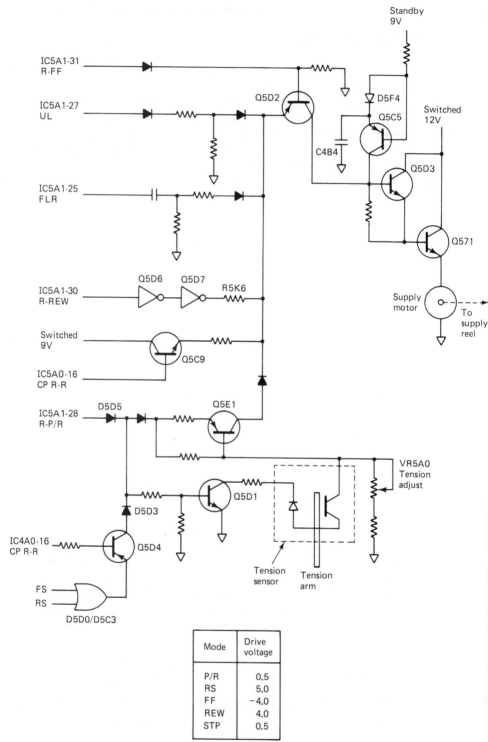

FIGURE 5-12 Supply motor drive circuits

drive through Q5D3/Q571 to the SU motor, until C4B4 discharges through the conducting Q5C5. This momentary drive to the SU motor brings the motor to a rapid stop, preventing slack tape.

5-2.13 Rewind

In rewind, the SU motor drives the supply reel at high speed and thus requires a high voltage (about 4 V). This high voltage is developed when a rewind command from IC5A1-30 is amplified and inverted by Q5D6/Q5D7 and is applied through R5K6 to the SU motor through Q5D2 and Q5D3/Q571.

5-2.14 Fast-forward brake

The Q5D6/Q5D7 combination also provides braking for the SU motor at termination of the FF mode. When FF is terminated, IC5A1 places a high on the R-REW line, enabling the rewind SU motor drive. Since the SU reel and motor are being pulled in reverse during FF, the SU motor drive acts as an electromagnetic braking force.

5-2.15 FF electrical brake release

As soon as the SU motor stops, the FF electrical brake release circuit shown in Fig. 5-13 directs IC5A1 to remove the high on the R-REW line, and the SU system comes to rest. Operation of the FF electrical brake release circuit is identical to that of the rewind electrical brake release circuit described in Sec. 5-2.9.

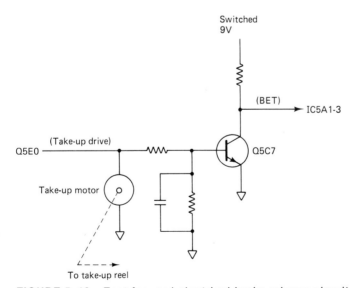

FIGURE 5-13 Fast-forward electrical brake-release circuit

5-2.16 Tape transport sensors

Sensors are used throughout the tape transport and mechanical sections of all VCRs to sense a variety of conditions. These conditions deal with protection of the tape or the VCR in the event of certain changes in operating conditions. Generally, the sensors act through the system control microprocessors to place the VCR in a stop condition. In most cases, the sensors place a high on certain control lines (or inputs to the microprocessors), any of which will stop the VCR. Figure 5-14 shows some typical sensor circuits.

End sensor. The end sensor circuit shown in Fig. 5-14 is used to prevent damage to the tape when the end of the tape is reached in play, record, fast forward, or forward search modes. During these forward modes Q574 does not conduct, since light originating from the cassette lamp (22, Fig. 5-4) is blocked from Q574 by opacity of the tape. However, the *transparent leader* just prior to the end of the (VHS) tape allows light from the cassette lamp to pass and thus turn on Q574. The turn-on of Q574 places a high at pin 7 of expander IC5A0.

Expander IC5A0, in conjunction with IC5A1, responds by generating the necessary logic to initiate a rewind mode (if the end of tape is reached during play or record) or to place the VCR in stop (if the end of tape is reached during FF). The programming of IC5A1 is such that if the end of tape is reached during a *timer record* operation (Chapter 14), rewind is not initiated. Instead, the VCR is placed in stop, precluding the possibility of recording over previously recorded material (in the event that the timer programming exceeds the capacity of the tape).

Start sensor. The start sensor shown in Fig. 5-14 protects the tape by placing the VCR in stop at the end of rewind and by prohibiting play and record if the tape is fully wound on the TU reel (a condition that could result in tape stretching). Normally, Q573 is turned off, since light from the cassette lamp is blocked by the tape. However, at the end of a rewind mode, or any other condition under which the tape is fully wound on the SU reel, the transparent leader at the beginning of the tape allows light from the cassette lamp to enable Q573. In turn, this turns on Q576, driving pin 8 of IC5A0 high, and places the VCR in stop (inhibiting both play and record modes).

Sensor stop. The sensor stop circuit shown in Fig. 5-14 prevents tape or VCR damage resulting from excessive slack tape or excessive moisture on the heads. In either case, the SST (sensor stop) line at ICTA0-11 is driven high, initiating the stop mode. In a VCR temporary power-off condition, the SST line is held high by conducting transistor Q5A3. When the VCR is switched on, the high generated on the T-PWV (power-on command) from the timer processor (IC8A1-13) disables Q5A3 and removes the standby 9-V supply from the SST line so that the VCR is not held in the stop mode.

Typical Front-Load Mechanical Operation   143

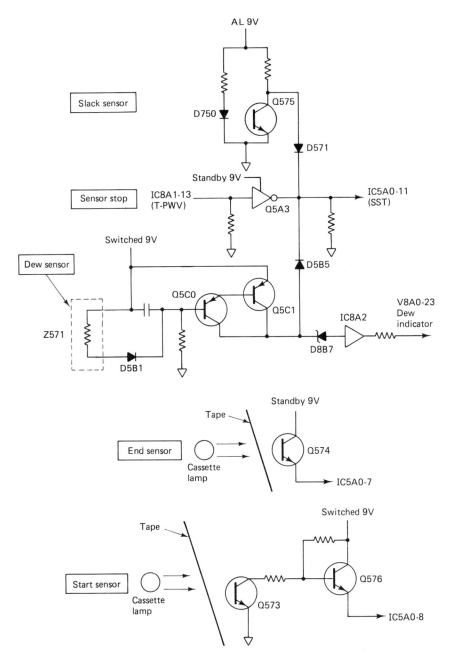

FIGURE 5-14   Typical tape-transport sensor circuits

Dew sensor. Should excessive moisture exist on or around the video heads, the moisture is detected by the dew sensor Z571. The resistance of Z571 varies in direct proportion to the amount of moisture detected. If the level of moisture increases, the resistance of Z571 also increases, lowering the base voltage of Q5C0, eventually turning on Q5C0. When Q5C0 turns on, Q5C1 is also turned on, driving the SST line high through D5B5 and holding the VCR in stop. The high produced by Q5C1 also exceeds the breakdown point of zener D8B7, is increased in amplitude by IC8A2, and is applied to the display V8A0, turning on the front-panel dew indicator.

Slack tape sensor. D570 and Q575 comprise a slack tape sensor which initiates the stop mode if tape slack becomes excessive. Normally, Q575 is conducting due to the reception of light emitted by LED D570. If the tape becomes slack, the light path is blocked, Q575 is turned off, and the SST line is driven high via D571. This places the VCR in stop.

5-3 MECHANICAL ADJUSTMENTS

We do not describe mechanical adjustments in this book for several reasons. First, the adjustment procedures are usually very complex, consuming considerable space. More important, the mechanical adjustments can apply to only one specific VCR. There are virtually no "universal" or "typical" mechanical adjustments for modern VCRs. This is in contrast to early-model VCRs, where the mechanical sections are more standard. Of course, you must follow the procedures outlined in the service literature, using the specified tools and fixtures, for any VCR. To do otherwise is a form of tampering.

However, we do cover troubleshooting/repair of those *circuits* associated with the mechanical section. The remainder of this chapter is devoted to such troubleshooting and repair procedures, using typical VCR circuits as examples.

5-4 TAPE LOADING MOTOR AND REEL MOTOR DRIVE CIRCUITS

In this section we discusss operation of typical tape loading and reel motor circuits. Note that the circuits are for a top-loading VCR.

5-4.1 Tape loading motor operation

Figure 5-15 shows the loading motor circuits for a typical top-load VCR (that shown in Fig. 1-1c). The loading motor operation is controlled by the system control microprocessor-B IC802 at pins 50 and 51, which are at high and low,

Tape Loading Motor and Reel Motor Drive Circuits   145

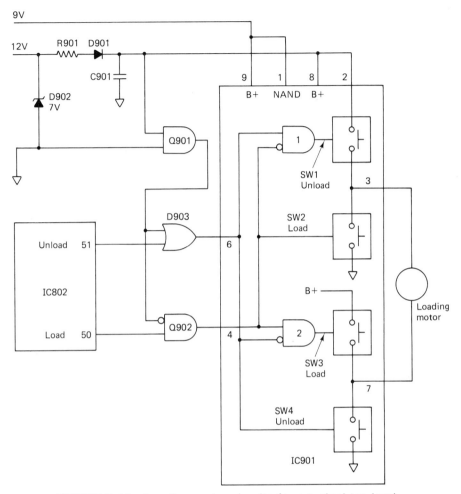

FIGURE 5-15 Loading motor circuits for a typical top-load VCR

respectively, during load. With the proper level of 12 V applied to the power sense circuit (D902, D901, Q901), the output of Q901 is low. This low is passed to the inverting input of Q902. The noninverting input of Q902 is at high (output from IC802-50). These two inputs produce a high at the output of Q902. This high is applied to the loading motor drive IC901-4. The low from IC802-51 is passed to one input of D903, while the other input of D903 is routed from the output of the power monitoring system Q901 which is at low. These two inputs produce a low at IC901-6. With a low at pin 6 and a high at pin 4, switches 2 and 3 within IC901 are on. This applies the proper voltage across the loading motor to rotate the motor in the load direction.

During unload, IC802-50 goes low and IC802-51 goes high. This produces a high at IC901-6 and a low at IC901-4, and causes switches 1 and 4 to close. With switches 1 and 4 closed, the loading motor is driven in the unload direction.

During play operation, when the tape is properly loaded into the VCR, if power is lost, the inverting input of Q901 goes low, but the noninverting input stays high due to the charge on C901. The output of Q901 then goes high. The high output of Q901 is routed through D903 to cause a high at IC901-6, and to Q902, producing a low at IC901-4. This places IC901 in the unloading mode until C901 discharges, and provides a partial unloading of the tape (to relieve tape tension).

The complete logic output of system control microprocessor-B IC802 and loading motor drive IC901 is shown in Fig. 5-16.

Tape loading motor troubleshooting/repair. If the VCR does not load or unload properly, first check the outputs of IC802 at pins 50 and 51. Compare these outputs to those shown in Fig. 5-16. If the outputs are absent or abnormal, suspect IC802. However, before you replace IC802, check the tape transport sensors (Sec. 5-2.16). It is possible that one of the sensors has been turned on (properly or improperly) and IC802 is prevented from operating properly. Next check the command lines between the operating controls and IC802. In most modern VCRs, the operating controls signals are passed through another system control IC before they are applied to the tape transport IC. In the case of the Fig. 5-15 circuit, trouble sensor signals are applied directly to IC802, but operating control signals are routed through system control microprocessor-A IC801.

If the outputs from IC802 are normal for the operating mode selected, then check the input logic at pins 4 and 6 of IC901. If these inputs are incorrect, suspect Q901, Q902, and D903. If the inputs are correct, check the output of IC901 at pins 3 and 7. If the outputs are incorrect, suspect IC901. If the outputs are correct, the problem is most likely in the load motor.

Motor condition	IC802		IC901			
	50	51	4	6	3	7
Stop	L	L	L	L	—	—
Brake	H	H	H	H	L	L
Loading	H	L	H	L	L	H
Unloading	L	H	L	H	H	L

FIGURE 5-16 Loading motor-drive I/O table

5-4.2 Reel motor drive operation

Figure 5-17 shows the reel motor circuits for a typical top-load VCR (that shown in Fig. 1-1c). System control microprocessor-B IC802 develops four control signals to determine the direction and speed of the reel motor operation. The control outputs are as follows:

The fast forward/rewind output at IC802-52 goes low when the VCR is placed in the fast-forward or rewind mode. This low is inverted by Q814 and

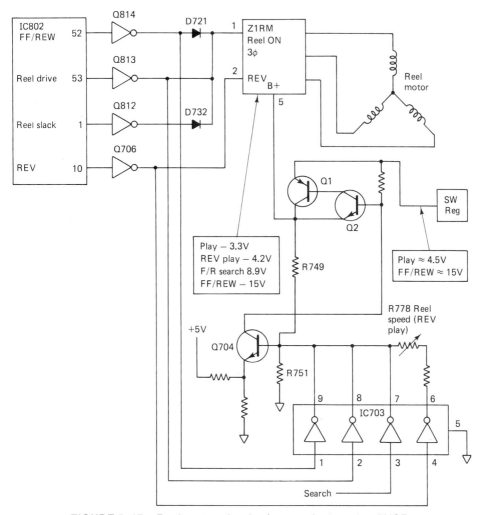

FIGURE 5-17 Reel-motor circuits for a typical top-load VCR

passed through D721 to pin 1 of Z1RM, the reel driver IC. The logic high at pin 1 causes Z1RM to apply power to the reel motor.

The reel drive signal at IC802-53 goes low when the VCR is placed in the play, reverse play, search forward, search reverse, video dub, or audio dub mode. The low is inverted to a high by Q813 and applied to Z1RM-1, turning on the reel motor.

The reel slack line at IC802-1 goes low during unloading mode. This low is inverted to a high by Q812 and applied through D732 to Z1RM-1, turning on the reel motor and taking up tape slack during unload.

The direction of reel motor rotation is determined by the signal at IC802-10. This output is low in reverse play, search forward, or rewind mode. The low is inverted to a high by Q706 and applied to Z1RM-2, causing the motor to rotate in reverse. The output at IC802-10 is high in other modes, causing the reel motor to rotate forward.

Reel motor speed is set by the amount of B+ applied at Z1RM-1. The B+ is taken from a programmable voltage regulator consisting of Q1, Q2, Q704, and IC703. Three of the four control lines from IC802 are applied to the inputs of IC703, along with the search signal. During various modes of operation, the inputs go high, turning on transistors within IC703 and switching in various resistances from the base of Q704 to ground. This changes the bias at the base of Q2 and determines the B+ voltage at Z1RM-5. An increase in B+ voltage produces an increase in reel motor speed, and vice versa.

The power source for the reel motor is taken from the switching regulator, which is part of the capstan power supply. The switching regulator output changes as necessary to accommodate the reel motor load changes.

Reel motor drive troubleshooting/repair. If the reel motor does not operate properly, first check for a high at Z1RM-1. If the high is missing, suspect defective system control circuits (IC802, Q812, Q813, Q814) and diodes D721, D73. If the high is present at Z1RM-1 check for B+ at Z1RM-5. If normal, suspect Z1RM. If B+ is missing, check for B+ applied to Q1/Q2 from the switching regulator. If B+ is missing from Q1/Q2, suspect the switching regulator.

If all the voltages and signals appear to be normal, but reel motor operation is not normal (fast or slow), try correcting the condition by adjustment of R778. Select the reverse play mode, monitor Z1RM-5, and adjust R778 for the correct voltage as shown in Fig. 5-17 or the service literature. While in reverse play mode, check that IC802-10 goes low and that the low is inverted to a high by Q706 for application to Z1RM-2.

If the reel motor operates normally in some modes but not in others, suspect IC703 and possibly Q706.

5-5 CASSETTE LOADING CIRCUITS

In this section we discuss operation of typical cassette load/unload/eject circuits. Note that the circuits are for a front-loading VCR.

5-5.1 Front-loading circuit operation

Figure 5-18 shows the major mechanical components for a typical front-load VCR (that shown in Fig. 1-1d). Figure 5-19 shows the related circuits. When power is first applied to the VCR, IC851 and IC852 are initialized and reset, placing the VCR in the stop mode. This causes the mode sense switch to apply a low to IC852-22 and a high to IC852-23. If there is no cassette in the VCR, the cassette-up switch S144 and the cassette-down switch S143 apply highs to IC852, pins 8 and 9, respectively. The takeup and supply end sensors are activated, applying a high to pins 26 and 27 of IC852. With these conditions met, IC852 awaits for a high to be applied to pin 7, indicating that a cassette is in the VCR and that power is on. Cassette-in switch S145 closes when the cassette is inserted into the holder. This applies a high to AND gate Q957. The other input to Q957 is a power-on signal from IC851. When both signals are present at the input of Q857, a high is applied to IC852-7, and cassette loading is ready to start.

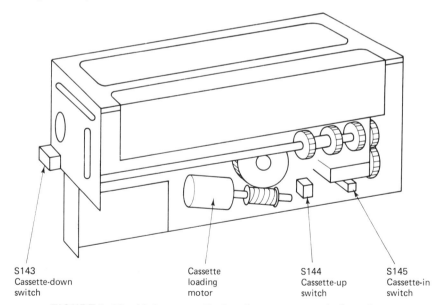

FIGURE 5-18 Major cassette-loading components for a typical front-load VCR

150   The Mechanical Sections

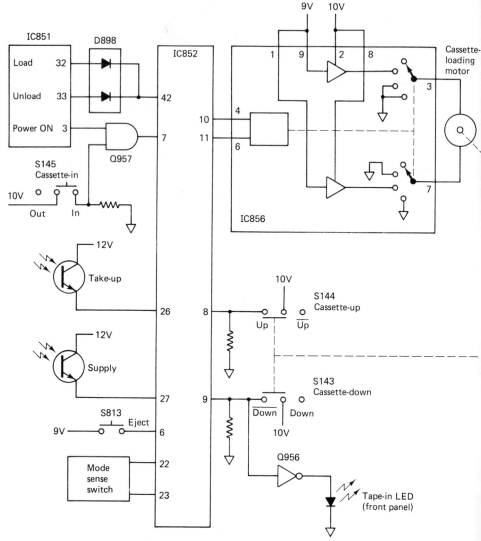

FIGURE 5-19   Front-load cassette-loading circuits

5-5.2   Front-load cassette loading sequence

The cassette loading-sequence timing is shown in Fig. 5-20. When power is applied and a cassette is inserted into the front of the VCR, IC852-7 goes high, causing a load signal to appear at IC852-10. This load signal causes IC856 to apply power (of the proper polarity) to the cassette loading motor. This pulls the cassette in and down. With the cassette loading started, the loading sequence must occur in a predetermined amount of time (6 to 8 seconds). Failure

Cassette Loading Circuits 151

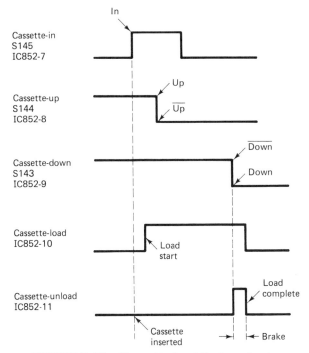

FIGURE 5-20 Cassette load timing chart

to complete the cassette loading sequence within that time causes IC852 to terminate the loading operation and eject the cassette.

With no cassette installed, S143 is in the $\overline{\text{DOWN}}$ position, and S144 is in UP. As the cassette is moved in and down over the reel hubs, S143 moves to DOWN, placing a low at IC852-9 and a low at the input of Q956. This low is inverted to a high by Q956 and turns on the tape-in LED on the VCR front panel. With a low at IC852-9, IC852-11 goes high, as does IC856-6. With a high at both pins 4 and 6 of IC856, a "brake" command is applied to the cassette loading motor, and the motor stops. By applying the signals to both the load and unload inputs of IC856, the motor is brought to an immediate stop and does not coast.

5-5.3 Front-load cassette unload/eject sequence

The cassette unload/eject sequence timing is shown in Fig. 5-21. When the eject switch S813 is pressed, a high is applied to IC852-6, and the cassette unloading sequence starts. If unloading is not complete within 6 to 8 seconds, IC852 terminates the unload/eject sequence and returns to the loading sequence (Sec. 5-5.2) to reload the cassette into the VCR.

When S813 is pressed, the high at IC852-6 causes IC852 to check that

152   The Mechanical Sections

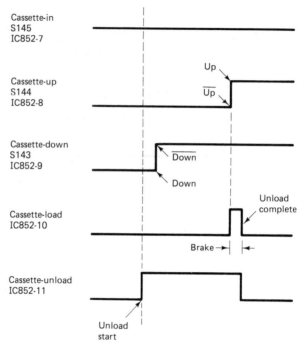

FIGURE 5-21 Cassette unload timing chart

the VCR is in stop (pins 22 and 23 of IC852 are low and high, respectively), and that neither tape load nor tape unload are active (IC852-42 low). With these conditions satisfied, IC852-11 applies a high to IC856-6, causing IC856 to drive the cassette loading motor in the unload/eject direction (up off the reel hubs and out toward the front of the VCR).

As the cassette is being unloaded, S143 returns from DOWN to $\overline{\text{DOWN}}$, making IC852-9 high. A short time later, S144 returns from $\overline{\text{UP}}$ to UP, making IC852-8 high. This combination causes IC852-10 to go high, applying a "brake" command to the cassette loading motor through IC856 and terminating the unload/eject sequence.

Front-load cassette unload/eject troubleshooting/repair.   Note that in the circuit of Fig. 5-19, the *tape end sensors* can cause an unusual failure symptom. If both tape end sensors receive enough light at the same time (from whatever cause), pins 26 and 27 of IC852 go high, and the mechanism will unload and eject, even in the middle of recording or playback. There are similar circuits on other VCRs. This is why still other VCRs use *infrared tape end sensors* that respond only to light from an infrared lamp within the VCR.

If either load or unload/eject appears abnormal, first check for the proper enable inputs to IC852. The VCR must (or should) be in the stop mode before the cassette load motor can operate (hopefully). This means that pins 22 and 43 should be low, while pin 23 of IC852 should be high. If the enable inputs are incorrect, check out the appropriate switches (mode sense), diodes (D869) and input sources (IC851). If the enable inputs are correct, then look for correct load and unload/eject commands at pins 7 and 6 of IC852, respectively. If these command inputs are incorrect, check out the appropriate switches (S145, S813), AND gate Q957, and input sources (IC851).

If the command inputs (and stop inputs) are normal, look for correct load and unload/eject signals at pins 10/11 of IC852 and 4/6 of IC856. If the signals are missing, suspect IC852. If the inputs at pins 4 and 6 of IC856 are correct, check the outputs at pins 3 and 7 of IC856. If the outputs are missing, suspect IC856. If the outputs at pins 3 and 7 are normal, but the cassette loading motor does not operate, suspect the motor.

If the cassette loading motor operates (and in the right direction) but the cassette appears to not reach the fully loaded or fully unloaded position, the problem is most likely in switches S143 and S144. The switches may be defective, or they may be improperly adjusted (improperly positioned along the cassette load/unload path, and not being actuated at the right time). Try manually actuating the switches (if you can reach them without jamming the mechanism). See if highs and lows are applied to pins 8 and 9 of IC852 when S143 and S144 are actuated. If not, the switches are probably bad. If the highs and lows from S143 and S144 are correct, but the motor does not operate at the right time, suspect IC852.

6

VIDEO HEAD CONFIGURATIONS AND SWITCHING CIRCUITS

This chapter describes the various configurations for video heads, and the related switching circuits. As discussed in Chapter 1, modern VCRs use four- and five-head systems instead of the two-head systems common in early-model VCRs. There are three general configurations (but many minor variations) for the video heads: four-head with 90° spacing, the 4X system, and five-head. Before we get to these three systems, let us consider some basic problems in VCR video heads.

6-1 VIDEO HEAD SWITCHING AND SELECTION CONCEPTS

The terms *head switching* and *head selection* are often interchanged in VCR literature. In this book, *head switching* refers to switching between two heads in a pair or set of heads, typically at a 30-Hz rate, using signals from the drum servo. *Head selection* refers to selecting a particular pair or set of heads to accommodate a given tape speed or playing time, such as SP, LP, and SLP (also known as 2H, 4H, and 6H). Head selection is controlled by relays (or analog switches) and is used only on VCRs with more than two video heads. Head switching occurs on all VCRs, but not necessarily in the same way. We will talk about basic head switching first.

Video Head Switching and Selection Concepts   155

6-1.1 Basic head switching

In order to prevent crosstalk between adjacent recorded video tracks during playback, VCRs use video heads with dissimilar azimuths ($\pm 7°$ for Beta, $\pm 6°$ for VHS). A particular video head, when scanning a track recorded by a head having the opposite azimuth, produces little or no output. This generates *noise in the picture.* Such noise, caused by slight errors in the lateral positioning of the tape with respect to the video heads, is called a *tracking error.* Such tracking errors are corrected by the *capstan phase control system,* using the control (CTL) pulses placed on the tape during record.

The tracking system (part of the capstan servo) adjusts the lateral position of the tape so that, as a head with a given azimuth is in a position to scan the tape, a track recorded by a head with the *same azimuth* is positioned to be scanned. The head is also made to scan through the center of the recorded track, with little or no overlap by the tracking system. In the simplest of terms, the user-operated tracking control shifts the phase of the capstan servo as necessary to produce proper tracking and prevent or minimize picture noise.

In addition to keeping the heads scanning the center of the track with the right azimuth, it is necessary that each head *scan the entire recorded track,* and to *maintain consistency* between tracks. This is the job of the *drum phase control system* (Chapter 7). The drum servo provides a highly stable reference frequency, phase-locked to the vertical sync frequency, and appearing at the vertical frame rate of the TV (30 Hz). This is the rate at which each individual video head is required to record, or play back, one track (or one vertical field) of information. Since two heads are used, two vertical fields are produced for each complete drum rotation. This produces 60 vertical fields, or 30 complete frames, each second. So each head contributes one-half of each frame, and each individual head input or output is selected 30 times per second.

As shown in Fig. 6-1, one complete drum FF cycle, occurring at a 30-

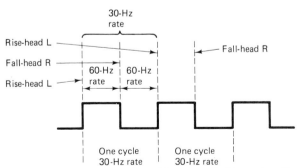

FIGURE 6-1 Conventional head-switching circuits

Hz rate, consists of a rise and fall of the FF pulse. Note that head L(left) is selected by the rise of each FF pulse, and head R(right) is selected by the fall of each pulse. So each head is selected alternately during each complete FF cycle, at twice the frame rate, or 60 Hz (which is the TV vertical field rate). Viewed individually, each head is selected only once for each FF cycle. If consideration is given only to R heads, or L heads, selection occurs at a 30-Hz rate. As a result, the head switching process yields one vertical field per head, or one complete vertical frame, when the head pairs are combined.

6-1.2 Basic head selection

Early-model VHS VCRs use two video heads, each with identical gap widths (typically 30 microns). Each head writes its own distinctive pattern on the tape. Since the VHS standard specifies a track width of 58 μ in the 2H mode, and each video head produces a 30-μ track, a *guard band* is produced between recorded tracks. Guard bands contain no recorded information, so the head produces no output when passing over a guard band.

Later two-head VCRs evolved to the *dissimilar-head concept*. Not only are the azimuths different, but each head has a different *gap width* (typically 26 and 31 microns, or 28 and 32 microns). This makes it possible to record in the 6H mode (special long play, extra long play, etc.), without guard bands, and to have still or freeze-frame pictures (sometimes referred to as *special effects* or *trick play*). With the tape moving slower in 6H, but the heads at the same speed, there are no guard bands (or the guard bands are very narrow). Either way, if the tape is stopped in 6H, each head scans at least *some* recorded information. The tape can be aligned so that, even with no tape movement (still or freeze frame), each head scans a portion of the appropriate recorded track. This produces an output for each video head and an acceptable still picture in 6H (minimum noise).

Unfortunately, still pictures in 2H are a problem using any two-head system. If tape movement is stopped on a 2H recording (with guard bands), the heads typically scan one track of information (with an output) or a guard band (without an output), producing noise in the picture. The problem of noisy still pictures in 2H is overcome by the *four-head concept*.

With one four-head system, two heads are used for 4H/6H speeds, and another two heads are used for 2H speed. The gaps for the extra pair of 2H heads are much wider. The video tracks recorded by the wider heads overlap, since the lateral speed of 2H is always 58 μ. When tape movement is stopped (still or freeze frame) in the wider 2H recording, even if there is some guard band, the wider heads never scan a guard band alone. Instead, the heads always scan some material recorded with the same azimuth, producing an output and a minimum of noise.

Note that in some four-head systems, all three speeds (2H, 4H, and 6H) are recorded with one pair of heads, but trick play is performed by the other pair of heads. Such a system is described in Sec. 6-2. For now, let us concern ourselves with head selection.

6-1.3 Basic four-head selection circuit

Figure 6-2 shows the video head selection circuits for a typical four-head VCR. Since all four heads are not used simultaneously, a means is provided to select

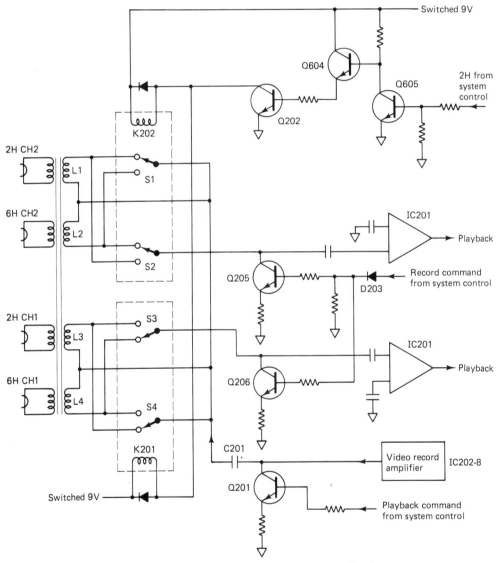

FIGURE 6-2 Basic four-head selection circuit

158   Video Head Configurations and Switching Circuits

a pair of heads to be used at a *particular speed*. In the Fig. 6-2 circuit, a 70/90 μ combination is used in 2H, and a 26/31 μ combination is used in 4H/6H. Head selection is done by two relays, K201 and K202, which are operated by signals from a speed sense or speed discriminator circuit through Q202/Q604/Q605. Outputs from the selected heads are applied to the head amplifier circuit IC201 during playback. During record, the output from the video record amplifier is applied to the selected heads through C201.

2H mode. In 2H, Q605 is turned on by a 2H command signal from system control. This disables Q604/Q202, and deenergizes K201/K202. Switches S1/S2 of K202 move to the lower position; switches S3/S4 of K201 assume the upper position. Head windings L2/L4 are shorted, the 6H heads are cut off, and only the 2H heads are used.

In 2H playback, the returns for coils L1/L3 are provided through C201 and Q201 (turned on by a playback command from system control). The outputs of the 2H heads are applied through L1/L3 to the channel-1 and channel-2 amplifiers of IC201.

In 2H record, Q201 is turned off by the absence of a playback command. This permits the output of the record amplifier to be applied through C201 and L1/L3 to the 2H heads. One side of the L1/L3 coils is returned to ground through Q205/Q206, which are turned on when a record command from the system control is applied through D203.

4H/6H modes. In 4H/6H, the 2H command applied to Q605 goes low, turning off Q605 and turning on Q604/Q202. This energizes relays K201/K202. Switches S1/S2 of K202 move to the upper position; switches S3/S4 move to the lower position. This shorts L1/L3, disables the 2H heads, and connects the 6H heads. Once the 6H heads are selected, the remaining circuits operate as described for the 2H mode.

6-2 FOUR-HEAD WITH 90° SPACING

Figure 6-3 shows the video head control circuits for a typical four-head VCR (that shown in Fig. 1-1d). Figure 6-4 shows the head configuration. Note that this head configuration is similar, but definitely not identical, to that described thus far. In the configuration of Figs. 6-3 and 6-4, the channel-1 and channel-2 (ch-1 and ch-2) heads are used during *normal* playback/record in *all three speeds*. In trick play (still, double speed, search), a second pair of heads, both designated ch-2', are used. The trick-play ch-2' heads are located 90° apart from the normal ch-1 and ch-2 heads.

As shown in Fig. 6-4, the bottom edge of the ch-2' heads uses the same reference as the normal ch-1 and ch-2 heads. This concept is different from other video head arrangements where the center of the heads is used as a ref-

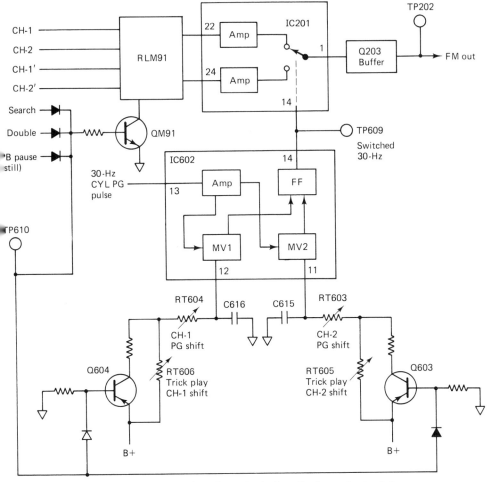

FIGURE 6-3 Video-head control circuits for a typical four-head VCR

erence. In order for the concept to work, the circuit conditions are modified during trick play to switch in the ch-2' heads and to position the heads to cross the normal channel-2 track. The positioning is done by *phase shift* of the cylinder servo signal used to switch the heads. The azimuth for the ch-2' heads is identical to the normal ch-2 head. This produces a good output when the ch-2' heads pick up the ch-2 recording. If the ch-2' heads should accidentally pass over the normal ch-1 track, the azimuths are reversed and there is no output (resulting in noise).

Head-switching relay RLM91 is energized by turning on QM91 during trick play with a high from system control (for search, double speed, still, or pause). With RLM91 turned on, the ch-2' signal is passed to pins 22 and 24

160 Video Head Configurations and Switching Circuits

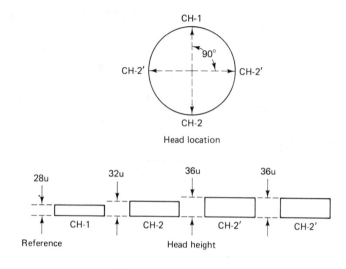

Mode	Head	Head thickness	Azimuth
SP(2), LP(4), SLP(6) Record or play	CH-1	28u	−
	CH-2	32u	+
Trick play	CH-2'	36u	+
	CH-2'	36u	+

Head comparison

FIGURE 6-4 Video-head configuration with 90° spacing

of head amplifier IC201. The switched 30-Hz signal (at IC201-14) that causes IC201 to switch between the ch-2' heads is produced by IC602. The IC201-14 is signal determines which FM head signal is amplified and applied to buffer Q203 through IC201-7. The switched signal can be monitored at TP609, while the combined FM (resembling that of Fig. 1-19) can be monitored at TP202.

The 30-Hz head-switching signal from IC602-14 is developed by signals received from the cylinder motor (Chapter 7) PG generator input at IC602-13. The signal is processed by two multivibrators with IC602. The time constant of the multivibrators is controlled by an RC time constant at pins 11 and 12 of IC602. By adjusting B+ at pins 11 and 12 (via RC time constants), the *delay time* that occurs between the occurrence of the cylinder PG pulse (also known as the drum FF pulse), and the rising and falling edges of the 30-Hz head-switching signal (Fig. 6-1) is controlled.

During normal playback, the inputs of Q603/Q604 are low, turning on both transistors. This applies B+ directly to the shifter controls RT603 and RT604, allowing for adjustment of the delay time to accommodate normal playback. During any one of the three trick-play modes, Q603/Q604 are turned off by a high at the inputs. This allows trick-play shifter controls RT605 and RT606 to be connected in series with the existing normal playback controls (RT603/RT604). The additional resistance changes the time-constant factor of each multivibrator, causing the *relative time shift* of 90° to occur in the 30-Hz head-switching signal. With the combination of RLM91 being energized and the 30-Hz switching signal being shifted, the FM output appearing at TP202 contains the signal from the appropriate pair of video heads for the particular mode of operation selected.

Four-head with 90° spacing troubleshooting/repair. If you get noisy video playback, check for proper logic signals applied at QM91, Q603, and Q604. The logic should be low for normal play and high for trick play. If the signals are absent or abnormal, suspect system control (Chapter 14). If the signals are correct, check that QM91, Q603, and Q604 are responding properly. (QM91 should turn on, and Q603/Q604 should turn off, during trick play.) If it becomes necessary to replace Q603/Q604, readjust all four shifter controls RT603, RT604, RT605, and RT606 as described in the service literature.

If you get no video playback, or the playback appears to be choppy, check for an FM signal at TP202 and a 30-Hz head-switching signal at IC602-13 and IC602-14 (or TP609 if more convenient). If the head-switching signal is absent or abnormal at IC602-13, suspect the cylinder servo (Chapter 7). If the head-switching signal is good at IC602-13, but not at IC602-14 (or TP609), suspect IC602. If you get good head-switching signals at IC201-14, but the FM is absent or abnormal at IC201-1, suspect IC201. If you get good FM at IC201-1, but not at TP202, suspect buffer Q203.

You can also check for signals from the heads at IC201-22 and -24. However, video head signals are usually low and difficult to measure. If you get any inputs from RLM91, and the inputs are equal in amplitude, the heads and RLM91 are probably good.

6-3 4X VIDEO HEAD SYSTEM

Figure 6-5 shows the head configuration for the 4X system. Note that four heads are used and that spacing between individual heads forming an active head pair is 180°. However, an adjacent spacing configuration is used in 4X. That is, additional video heads R and L are positioned adjacent to primary video heads R' and L'. *The adjacent heads are located one horizontal scan line apart (about 12°).* (The letter designation of L or R denotes left or right azimuth. The prime notation after the letter designation—L' and R'—denotes the head pair used primarily in the 2H mode.)

Only two heads are used during normal playback/record at a given speed. Since the two heads of each pair retain conventional 180° spacing, head-

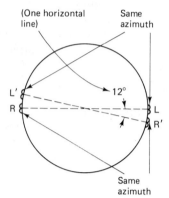

FIGURE 6-5 4X system video-head configuration

L = Left azimuth (6H)
R = Right azimuth (6H)
L' = Left azimuth (2H)
R' = Right azimuth (2H)

switching techniques described thus far remain essentially unchanged. The R' and L' video heads are used exclusively for 2H, while R and L heads are used for both 4H and 6H.

Due to the close proximity of the heads (12°), adjacent heads are contained in the same head assembly, as shown in Fig. 6-6. Also shown in Fig. 6-6 are the individual gap widths for each head. Note that the gap widths of the 2H heads (R' and L') are *smaller* than the gap widths used in many other four-head systems. With smaller gap widths, some guard bands are produced in 2H. However, the 4X system overcomes the potential noise problem by a unique combination or usage of the four heads. Figure 6-7 lists the head combinations used for the various special effects modes.

In *play/record* at the 2H speed, only the 2H heads are used; at 4H/6H speed only the 6H heads are used. *Speed search* (SS) uses all four heads, while 6H speed search uses only two heads (since there are no guard bands to make

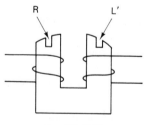

R = 32u (6H)
L = 30u (6H)
R' = 32u (2H)
L' = 45u (2H)

FIGURE 6-6 Head assembly gap widths for 4X

Speed \ Mode	PB/REC	Speed search	Still	Slow
2H	R'L'	R'L'RL	L'L	L'LR'
4H/6H	RL	RL	RR'	RR'L

FIGURE 6-7 Head usage for 4X

noise). In *slow motion, three heads* are used. In *still*, two heads are used, but *both heads* scan the same recorded track (or vertical field), as in the system of Sec. 6-2. This minimizes both horizontal and vertical jitter.

6-3.1 4X circuit operation

The circuits involved in the 4X video head system are very complex. Figure 6-8 shows the head selection and switching circuits for a VCR using the 4X system (that shown in Fig. 1-1a). The following is a brief summary of circuit operation.

Instead of a mechanical relay, the 4X system uses electronic analog switches for both head switching and selection. Conventional head switching is done by IC201 and IC203, which accept output signals from all four heads simultaneously. Head output selection is done by internal switches in IC2A2, sequenced by a drum FF pulse (called the cylinder PG pulse in other VCRs, just to confuse you).

IC201 accepts information from the 6H heads, and IC203 accepts information from the 2H heads. The drum FF switching pulse (30 Hz) is applied to IC201 and IC203 at pins 4 and 28, respectively, and controls the analog switches within the ICs. The drum FF signal switches the outputs at IC201-9 and IC203-2 back and forth between the right azimuth head and left azimuth head. Head outputs are then directed to IC2A2, which contains pairs of series analog switches.

The 6H signal from IC201-9 is applied to IC2A2-1, and the 2H head output from IC203-2 is directed to IC2A2-8. The analog switches within IC2A2 are controlled by the logic inputs at pins 5, 6, 12, and 13 of IC2A2. In the 6H mode, a high is applied to pins 5 and 13 of IC2A2, closing the 6H analog switches. The 6H head signal input at IC2A2-1 is passed to pin 4 of IC2A2 via the closed analog switches. In the 2H mode, pins 6 and 12 of IC2A2 are high, closing the 2H analog switches, and passing the 2H head signal input from IC2A2-8 to pin 11 of IC2A2. Note that the use of two analog switches in parallel minimizes the possibility of unwanted signals passing through and causing video noise.

The logic for the analog switches of IC2A2 is taken from the gate array IC4A7 and inverter transistors Q2A0 and Q2A1. IC4A7 determines the source logic controlling the selection of head outputs at a given speed and in a given mode of operation. The output of IC4A7 at pin 15 turns on the analog switches

164 Video Head Configurations and Switching Circuits

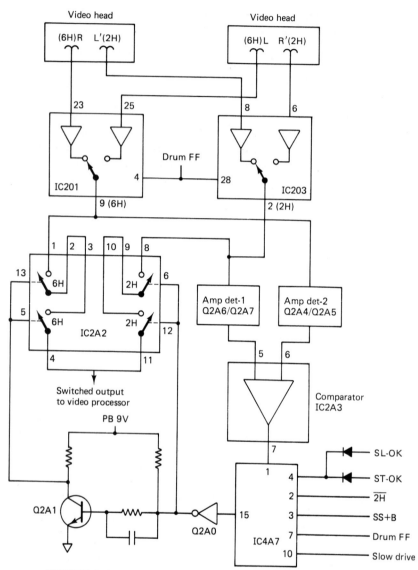

FIGURE 6-8 Head selection and switching circuits for 4X

in IC2A2. The output logic can be determined by a number of inputs, at pins 1, 2, 3, 4, 7, and 10 of IC4A7.

The input at pin 1 of IC4A7 is somewhat unique when compared to the remaining inputs. The IC4A7-1 input is turned on and off by the combination of logic at IC4A7-2 (from the automatic speed selector) and IC4A7-3 (from the speed-search circuits). The amplitude detectors supplying the input to IC4A7-1 via the comparator in IC2A3 are used only in 2H speed search.

In play/record, the logic at IC4A7-2 ($\overline{2H}$) informs IC4A7 of the tape speed mode. The $\overline{2H}$ line is high in the 4H/6H modes and low in the 2H mode. In 2H, IC4A7-15 goes low. This low is inverted to a high by Q2A0, and is applied to pins 6 and 12 of IC2A2, closing the 2H analog switches and connecting the 2H heads to the output at IC2A2-11. In 6H, the $\overline{2H}$ line and IC4A7-2 go high, producing a high at IC4A7-15. This high is inverted to a low by Q2A0, disabling the 2H analog switches. The low from Q2A0 is inverted again by Q2A1 and is applied to pins 5 and 13 of IC2A2, closing the 6H analog switches and connecting the 6H heads to the output at IC2A2-4.

4X system troubleshooting/repair. Although the circuits are far more complex, troubleshooting and repair for the 4X system are essentially the same as for the four-head system described in Sec. 6-2. That is, you check for proper control logic and head-switching signals and monitor the FM video output to the video processing circuits.

If the logic circuit signals are absent or abnormal, suspect the system control (Chapter 14). If the logic signals are good, look for problems in gate array IC4A7 and Q2A0/Q2A1. If head-switching signals are not correct, suspect the drum servo (Chapter 7).

If the logic and head-switching signals are all good, but there is considerable video noise, suspect IC2A2. If the video is choppy (as monitored at IC201-9 and/or IC203-2), suspect IC201 and IC203. Again, it is difficult to monitor the outputs directly from the video heads.

If you have problems (video noise, bad picture) only in the 2H mode, suspect Q2A4/Q2A5, Q2A6/Q2A7, IC2A3, and (of course) the 2H heads.

6-4 FIVE-HEAD SYSTEM

The fifth head in a five-head system is used primarily for slow motion or still (freeze-frame) operation. So we will concentrate on these functions in our discussion. Figure 6-9 shows the slow/still playback head selection circuits for a typical VCR (that shown in Fig. 1-1c). Figure 6-10 shows the relationship of the tracks, heads, and FM video output during still (freeze-frame) operation. Figure 6-11 shows the basic video head specifications for a five-head system.

As shown in these illustrations, the cylinder has five heads: two SP heads, two LP/SLP heads, and a fifth *still-field* head. The channel-1 and channel-2 SP heads are used in normal playback/record in the SP mode. In LP/SLP, the channel-1 LP/SLP head and the channel-2 LP/SLP head are used for normal playback/record. The SP heads are 75 μ thick, while the LP/SLP pair of heads are 28 μ thick. The fifth still-field head is 32 μ thick.

The fifth still-field head is located one horizontal line away from the channel-1 LP/SLP head, with the *same azimuth* as the channel-2 LP/SLP head. During field-still operation, the tape mechanism stops the tape so the

166 Video Head Configurations and Switching Circuits

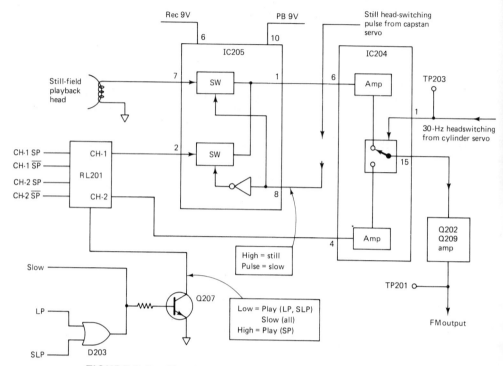

FIGURE 6-9 Slow/still playback head-selection circuits for five-head system

channel-2 LP/SLP head and the still-field head play back the *same video track*. By using the two video heads at approximately 180° apart, and at the same azimuth, playback of two identical fields per frame is obtained.

6-4.1 Slow/still playback head selection

During playback, either the LP/SLP pair of heads or the SP pair of heads, are selected by RL201 and transistor Q207. During slow motion, LP, or SLP modes, Q207 is turned on, energizing RL201. (Note that the LP/SLP pair of heads is used during slow motion, whether the material is recorded in SP or SLP.) The channel-2 video head output is passed directly to IC204-4, which is the input for one-half of the preamplifier circuit used to add the outputs of both heads. The channel-1 video head output is applied to IC205-2, while the still-field playback head output is applied to IC205-7. IC205 is a video switch used to select which head output appears at IC205-1 and IC204-6, as determined by the signal at IC205-8 from the capstan servo (Chapter 8).

When the signal at IC205-8 is high, the output from IC205-1 is taken from the still-field head. When IC205-8 is low, the signal at IC205-1 is from the channel-1 LP/SLP head. Either way, the two head signals at pins 4 and

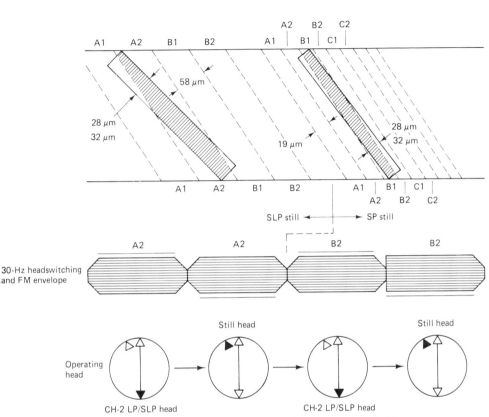

FIGURE 6-10 Relationship of tracks, heads, and FM video output during still (freeze-frame) operation

6 of IC204 are combined in IC204 to produce an FM output (resembling that of Fig. 1-19). IC204 is controlled by a 30-Hz head-switching signal (applied at IC204-1) from the cylinder servo (Chapter 7). The FM output from IC204-15 is amplified by Q202/Q209 and can be monitored at TP201.

During slow motion, the tape movement stops for a period of time and moves for a period of time. During the time that the tape is stopped or still (a high at IC205-8), the channel-2 LP/SLP head, and the still-field head are turned on to provide maximum video output. During the time that the tape is moving (a low at IC205-8), channel-1 and channel-2 LP/SLP head outputs are turned on.

Five-head slow/still troubleshooting/repair. If you get normal playback at all speeds, *but noise in the still-field mode,* first check that IC205-8 is high. If not, check the capstan servo (Chapter 8). If IC205-8 is high, then check

168 Video Head Configurations and Switching Circuits

Head name	CH-1 SP	CH-2 SP	CH-1 LP/SLP	Still play	CH-2 LP/SLP
Head thickness (μm)	75	75	28	32	28
Modes (record, play, search)	Exclusively for SP speed		For LP/SLP speeds		For LP/SLP speeds
Part of tape feed period in slow play, frame advance			For SP/LP/SLP speeds		For SP/LP/SLP speeds
Stop period in slow play, frame advance, and still play				For SP/LP/SLP speeds	For SP/LP/SLP speeds

FIGURE 6-11 Basic video-head specifications for five-head system

IC205-1 for the presence of still-field playback signals. (Note that these signals are low in amplitude and may be difficult to measure. However, you should be able to detect the presence of signals with a sensitive oscilloscope.) Compare the signals at IC205-1 (and IC204-6) with those at IC204-4. If the signals at IC205-1 are absent or abnormal, suspect IC205 or the still-field head.

If the signals at IC205-1 are normal, check for FM at TP201 and a 30-Hz head-switching signal at IC204-1. If the head-switching signal is absent or abnormal at IC204-1, suspect the cylinder servo (Chapter 7). If you get good head-switching signals at IC204-1, but the FM is absent or abnormal at IC204-15, suspect IC204. If you get good FM at IC204-15, but not at TP201, suspect Q202 or Q209.

If you get normal playback at all speeds, *but noise in the slow-motion mode,* first check for noise in still-field as described. If still-field is good, check that IC205-8 is pulsed low by the signal from the capstan servo (Chapter 8). If the low-pulse signal is absent or abnormal, suspect the capstan servo. If the signal at IC205-8 is normal, suspect IC205.

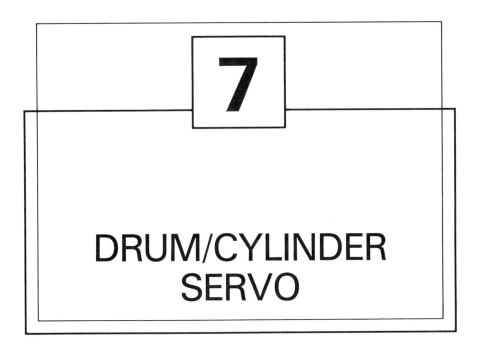

DRUM/CYLINDER SERVO

This chapter is devoted to troubleshooting and repair of VCR drum/cylinder servo circuits. (Note that the drum or cylinder servo may also be called the *head servo,* or *scanner servo,* in some VCR literature.) No matter what it is called, the servo must: (1) control the *speed* of rotation for the *videoheads,* (2) control the *phase* relationship of the *video heads* to the tape, and (3) provide drive for the *video head motor.*

The overall functions of VCR servo systems are described in Sec. 1-4. The following sections describe details and characteristics of video head servo circuits (drum or cylinder, whichever you prefer).

7-1 BASIC DRUM/CYLINDER PHASE CONTROL

Figure 7-1 is the overall block diagram of the cylinder phase control circuit for a typical VCR. Figure 7-2 shows the related waveforms.

As shown in Fig. 7-1, the 30-Hz cylinder tach pulse from a pulse generator installed in the lower part of the cylinder shaft is supplied to the comparison signal circuit for the cylinder in IC501 at pin 19. This cylinder tach pulse is used in both playback and recording. The pulse at IC501-19 is applied to the positive pulse amplifier and negative pulse amplifier, to be detected separately as positive and negative pulses. The pulse applied to the positive amplifier is detected as the positive pulse and amplified to a level suitable to trigger MM1 (monostable multivibrator 1), which acts as an adjustable delay circuit to determine the video head switching phase. The delay can be changed

170 Drum/Cylinder Servo

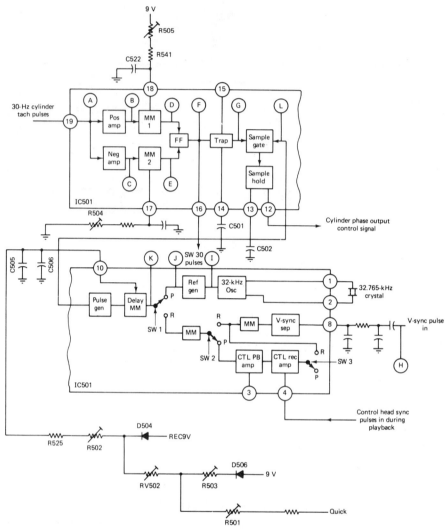

FIGURE 7-1 Cylinder phase-control circuit

by the time constant of C522 and R541 connected to IC501-18. Channel-1 switching phase is adjusted by R505. The pulse applied to the negative amplifier is delayed by MM2, as adjusted by R504.

The output from MM1 and MM2 is applied to an FF. MM1 is connected to the reset input of the FF, with MM2 connected to the set input. The pulse from the FF (waveform F of Fig. 7-2) is the video head switch pulse, called SW30, and supplied to the video luminance and color circuits. Pulse SW30 is adjusted by R504 and R505 to become a 30-Hz rectangular wave with a 50%

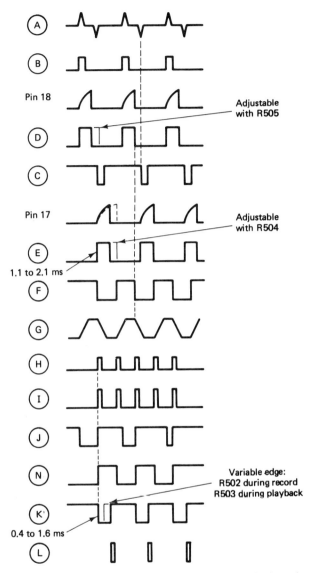

FIGURE 7-2 Waveforms associated with cylinder phase-control circuit

duty cycle. The output of the FF is converted to a trapezoidal waveform which is applied to a sample-and-hold circuit as a comparison signal. The sample-and-hold circuit also receives a reference signal, which is compared with the trapezoidal signal to produce the desired cylinder phase output control signal at IC501-12.

During record, switches SW1, SW2, and SW3 in IC501 are electrically connected to the record (R) position, and the broadcast V-sync signal is used as the reference signal. The V-sync component of the composite TV sync signal (supplied by the luminance circuit) is applied to IC501-8. The V-sync signal is separated by the V-sync separator and shaped by the MM to produce a 30-Hz rectangular wave. This pulse is amplified through the control pulse recording amplifier and supplied to the control head (through IC501-4), where the pulse is recorded on tape to become the capstan control signal (Chapter 8) during playback.

During playback, switches SW1, SW2, and SW3 are electrically connected to the playback (P) position, and the internal crystal-controlled signal is used as the reference signal. The playback reference signal is obtained by dividing the 32.765-MHz signal (developed by the crystal oscillator using an external crystal connected at pins 1 and 2 of IC501) by 1093. This 1093 division occurs in the reference generator circuit.

During either playback or record, the delay MM circuit receives a 30-Hz reference signal through SW1. During playback, the time constant of the delay MM is increased to provide correct tracking. Diode D504 is off during playback, and the delay MM time constant is approximately equal to the total of C505, C506, R502, R525, and RV502. Diode R504 is turned on during record by a 9-V control voltage, removing RV502 and R503 from the circuit and decreasing the time constant of the delay MM to the equivalent of C505, C506, R502, and R525. Variable resistor R502 is the record timing or phase control used to set the recorded V-sync signal at the correct position. R501 and R503 provide a similar function during playback (R501 for the quick mode and R503 for other modes).

The pulse generator differentiates the pulse obtained through the delay MM to produce the reference sampling pulse applied to the sample-and-hold (S/H) circuit. When the trapezoidal wave (comparison signal) and the sampling pulse (reference signal) are applied to the S/H circuit, an error voltage corresponding to the phase difference is generated and applied to the capstan speed control circuit (Chapter 8). Capacitor C502 holds the error voltage between samplings.

If the speed and phase of the cylinder are correct, the trapezoidal and sampling pulses line up as shown in Fig. 7-3. Notice that the sampling pulse lines up in the center of the trailing edge of the trapezoid. When the trapezoid waveform and the sampling pulse are supplied to the S/H stage, the voltage level of the trapezoid ramp is sampled (at the point where the sampling pulse intersects the ramp). This is given as X-volts in Fig. 7-3.

If the cylinder motor speed increases, the trapezoid wave phase leads with respect to the sampling pulse as shown in Fig. 7-4. The sampling position moves lower on the ramp and the error voltage decreases, making the cylinder motor rotate at a lower speed. If the cylinder motor speed decreases, the trapezoidal wave lags behind the sampling pulse as shown in Fig. 7-5, and the

Basic Drum/Cylinder Phase Control   173

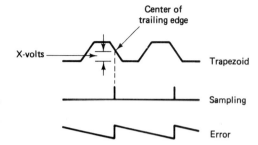

FIGURE 7-3 Relationship of pulses when cylinder speed is normal

sampling position moves higher on the ramp. As a result, the error voltage increases, making the cylinder rotate faster.

In summary, when the sampling pulse is used to sample the trapezoid ramp, a variable voltage results that is in direct relationship to the relative position of the sampling pulse on the ramp. Because this voltage represents video head position, the voltage can be used to control cylinder motor phase. However, the phase control developed by this circuit is limited to that which can be detected on the slope of the ramp, and is used only as a vernier speed

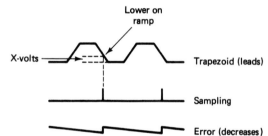

FIGURE 7-4 Relationship of pulses when cylinder speed increases

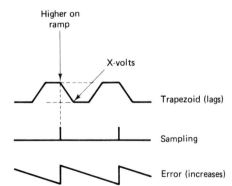

FIGURE 7-5 Relationship of pulses when cylinder speed decreases

control (or phasing control) voltage. The actual speed of the cylinder motor is controlled by the speed control circuit, discussed next.

7-2 BASIC DRUM/CYLINDER SPEED CONTROL

Figure 7-6 is the overall block diagram of the cylinder speed control circuit. Figure 7-7 shows the related waveforms.

As shown in Fig. 7-6, the 120-Hz signal output from the cylinder FG is applied to the FG amplifier through pin 6 of IC502 and C543. The FG amplifier has a differential input at terminals 5 and 6 of IC502, and a feedback output at IC502-8. A reference voltage applied to IC502-5 is adjusted by R506. The feedback signal is applied through R593 and C549. The output of the FG amplifier is applied to a delay circuit and to an S/H circuit, through doubler and former circuits. The signal is shaped by the former and converted to double frequency by the doubler. The delay circuit produces an approximate 20-μs delay.

The output from the delay circuit is used as the trigger for the sawtooth generator, which produces a sawtooth wave. The slope of the sawtooth waveform is determined by the time constant of the "C" and "R" connected to IC502 pins 11 and 9, respectively. Note that the resistance is made adjustable so that the speed can be set. As the sawtooth wave slope becomes sharper, speed increases.

The output of the sawtooth generator is applied to the S/H circuit, which also receives a delayed sampling pulse from the delay circuit. Capacitor C536 holds the voltage between samplings. The two inputs to the S/H circuit produce an error voltage output at pin 13 (waveform G in Fig. 7-7). The error voltage is applied to the cylinder motor drive (Sec. 7-3) through an adder, comparator, amplifier, and motor switch.

The sampling action is shown in Figs. 7-8, 7-9, and 7-10. As shown in Fig. 7-8, the sawtooth ramp is sampled at a fixed level determined by the delayed sample pulse amplitude. Since the lag is constant, when motor speed increases, the sawtooth ramp is steeper, the sampling position moves lower on the ramp, and the error voltage is reduced, as shown in Fig. 7-9. This reduces motor speed. When motor speed is reduced, the sawtooth ramp is less steep, the sampling position moves higher on the ramp, and the error voltage is increased, as shown in Fig. 7-10. This increases motor speed.

The error signal at pin 13 of IC502 is passed through a low-pass filter composed of R560 and C535 (to remove the sampling frequency component) and applied to the adder through IC502-14. The adder also receives an error signal from the phase control circuit (Sec. 7-1). The two error signals are added and applied to the motor drive at IC502-3 as the speed control signal (Sec. 7-3).

The comparator compares the sum of the speed control, phase control,

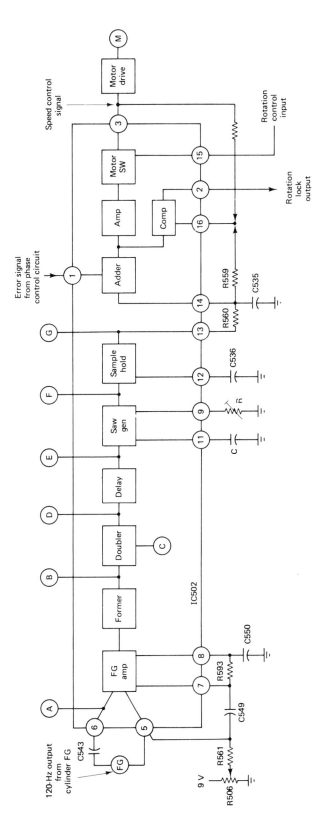

FIGURE 7-6 Cylinder speed-control circuit

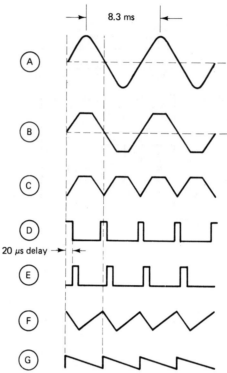

FIGURE 7-7 Waveforms associated with cylinder speed-control circuit

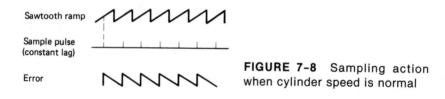

FIGURE 7-8 Sampling action when cylinder speed is normal

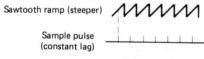

FIGURE 7-9 Sampling action when cylinder speed increases

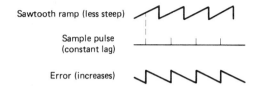

FIGURE 7-10 Sampling action when cylinder speed decreases

and reference voltage outputs. If the motor slows down for any reason to a point where the servo cannot control the speed properly, pin 2 of IC502 goes high and the motor stops (through action of the system control circuits, Chapter 14). IC502-2 is normally low.

7-3 BASIC DRUM/CYLINDER MOTOR DRIVE

Figure 7-11 shows the basic motor drive circuit for most modern VCRs. As shown, the motor is three-phase. Figure 7-12 shows the position of the rotor, stator coils, and Hall elements devices for the cylinder motor. (Hall elements operate on the Hall-effect principle and produce a current that is controlled by the magnetic field surrounding the element. When the magnetic field is alternating, the output current from the Hall element alternates.)

As shown in both Figs. 7-11 and 7-13 (drive circuit and related waveforms), the motor drive circuit for each of the three phases is composed of a Hall element device, a motor predrive, and a driver. When the motor turns, the polarity of the magnetic field applied to the Hall element alternates (because of rotation by the rotor magnet). This produces a three-phase sine wave, with phases U, V, and W obtained from the related Hall element. The three-phase sine wave is selected and amplified by the predriver, and the resultant current is applied to the motor armature by the driver.

Note that the cylinder motor does not use the motor-reverse provision (at pin 6 as shown in Fig. 7-11). The IC shown is a universal device used both for cylinder and capstan motors (which do require a motor-reverse feature).

As shown in Fig. 7-13, the cylinder motor has a three-section Hall element device. The signal obtained from this device is actually a six-phase sine wave. The predriver selects three pairs of Hall element outputs and converts the pairs to three pairs of controlled outputs which are applied to transistors within the driver. Transistors QU1, QV1, and QW1 are active-low and control the current applied to the corresponding armature windings of the motor. Transistors QU2, QV2, and QW2 are active-high and ground the corresponding windings.

Motor speed depends on the current supplied to the windings. In turn, the amount of current depends on the amplitude of the speed control signal

178 Drum/Cylinder Servo

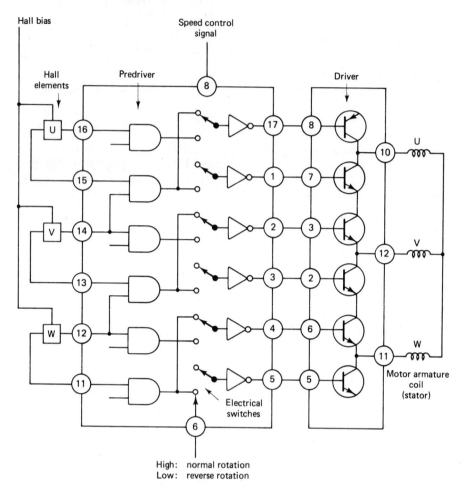

FIGURE 7-11 Basic servo drive circuit

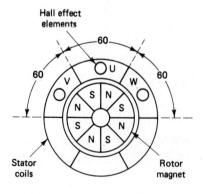

FIGURE 7-12 Positions of rotor, stator coils, and Hall-element devices of cylinder motor

Basic Drum/Cylinder Motor Drive 179

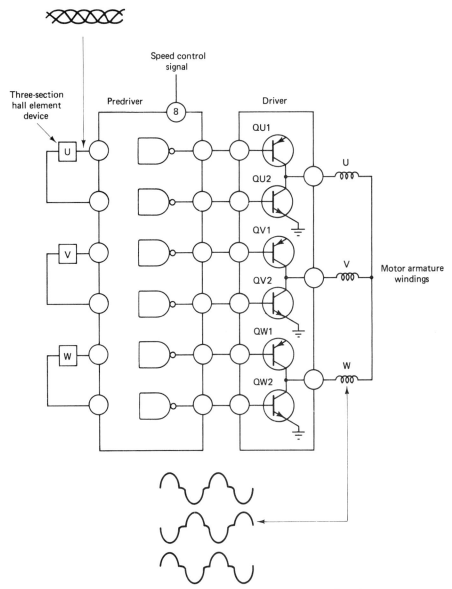

FIGURE 7-13 Cylinder motor drive circuit and related waveforms

(Sec. 7-2) applied to IC502-8. The direction of rotation for the cylinder motor is always the same, and only motor speed (or phase) is subject to control. (In the case of the capstan motor, Chapter 8, the direction of rotation is also controlled, as determined by logic from the system control applied to pin 6 of the motor drive IC.)

7-4 TYPICAL CYLINDER SERVO OPERATION

Figure 7-14 shows the overall block diagram for the cylinder servo system of a typical VCR (that shown in Fig. 1-1c). Figure 7-15 is the block diagram for the phase and speed loops of the cylinder servo. Note that the phase-loop circuitry is contained within IC501, while the speed loop portion is within IC502. The outputs of the phase and speed loops are added together to determine drive voltage to the cylinder motor drive IC.

During special-effect operation or the search mode, an added correction is applied to the servo control voltage. A search correction voltage is applied through IC503, as described in Sec. 7-4.2. Another correction voltage is applied through IC504 as part of the capstan servo operation (discussed in Chapter 8).

7-4.1 Cylinder phase and speed loop operation

The phase-loop portion of the cylinder servo is a PWM (pulse width modulated) system contained within IC501.

During playback, the reference signal for the phase loop is derived from a 3.58-MHz signal at IC501-25. This is divided down by circuits within IC501 to 30 Hz. The comparison signal is derived from the 30-Hz cylinder PG signal input at IC501-14. This signal is delayed by two MVs. The time constant of the two MVs is controlled by the two PG shift adjustments at pins 12 and 13 of IC501. One output from the MVs is the 30-Hz head-switching signal (Chapter 6) at IC501-15. The other output from the MVs is applied to the PWM

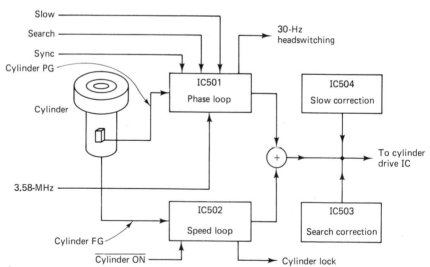

FIGURE 7-14 Overall block diagram for the cylinder servo system

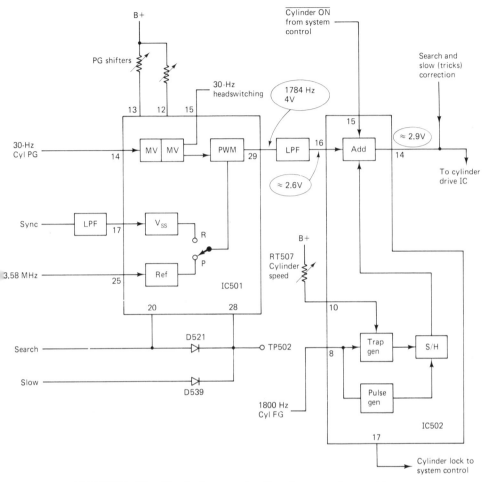

FIGURE 7-15 Block diagram for the phase and speed loops of the cylinder servo

circuit, which also receives the 30-Hz record or playback reference signal. The two signals are compared in the PWM circuit to determine the correction necessary for proper phase.

During record, the reference signal is the divided-down vertical sync signal at IC501-17 (instead of the 3.58-MHz signal). The output of the PWM at IC501-29 is a 1784-Hz signal, which is passed through a low-pass filter (to remove the ac component). The resultant dc output (about 2.6 V) from the low-pass filter is applied to pin 16 of the speed-loop IC502.

During the search or slow modes, the system control (Chapter 14) circuits apply a high at IC501-28. This high, which can be monitored at TP502, places the output of the PWM system in a fixed 50% duty cycle, inhibiting any phase shift of the cylinder motor.

IC502 is essentially the coarse speed control for the cylinder servo system; it receives a primary input from the cylinder 1800-Hz FG signal at IC502-8. The signal applied to pin 16 of IC502 from the phase-loop IC501 is a form of vernier speed (or phase) control voltage.

The speed-loop circuits within IC502 are of the sample-and-hold type. The 1800-Hz cylinder FG signal at IC502-8 is processed and passed to the sample-and-hold circuit, which develops a correction voltage. This voltage is passed to the adder circuit, along with the phase-control voltage. The adder circuit passes the combined control voltage to the cylinder motor drive IC through IC502-14.

Note that the cylinder motor drive signal is active only when system control applies a cylinder-on signal (a low at IC502-15). This enables the adder circuit to output the motor drive voltage. Also applied to the motor drive is the search correction voltage (Sec. 7-4.2) and the slow correction voltage (Chapter 8).

During initial start-up of the cylinder servo system, the cylinder motor is at zero rotation. So when the cylinder-on command occurs, the cylinder-lock signal at IC502-17 goes high for a short period of time (until rotation of the cylinder motor reaches the proper value), and then returns to low. The cylinder lock tells the system control circuits that the cylinder motor is not up to speed. System control then places the VCR in the stop mode. However, system control does not monitor the cylinder-lock signal during load and initial start-up. After proper loading, and when the VCR is in play mode, system control then monitors the cylinder-lock signal for proper level. If the signal is low, the VCR remains in play. If high, system control places the VCR in stop.

Cylinder phase and speed loop troubleshooting/repair. If the picture is out of horizontal sync (the most common symptom for trouble in the cylinder phase/speed circuits), play back a known good tape, or a test tape (Chapter 2). Connect a +5-V source to TP502, locking the PWM output of IC501 to the 50% duty cycle. Check for the presence of a signal at IC501-29. The signal should be about 4 V at 1784 Hz, with a 50% duty cycle. If the signal is absent or abnormal, suspect IC501. If the signal is correct, look for approximately 2.6-V dc at IC502-16. If this voltage is absent or abnormal, suspect the low-pass filter network. If the signal is normal at IC502-16, check for a low at IC502-15. If IC502-15 is high, indicating that the cylinder is locked, check the system control circuits. If IC502-15 is low, and there is approximately 2.6 V at IC502-16, try monitoring the dc voltage at IC502-14 while adjusting the cylinder speed control RT507. Adjust RT507 for approximately 2.9 V at IC502-14 and check that the TV picture is in horizontal sync. (However, if the phase control IC501 is disabled by +5 V at pin 28, there may be some noise bars floating through the picture, even if the picture is in sync. Remove the phase-disabling +5 V from IC501-28, TP502, and recheck for noise bars and horizontal sync.) If you cannot get proper horizontal sync by adjustment of RT507, with all inputs to IC502 correct, suspect IC502.

7-4.2 Cylinder servo search correction

During search operation, tape speed is increased by 5 in SP or 15 in SLP. Due to this increase in tape speed, the cylinder rotation speed must also be changed slightly to maintain proper horizontal sync frequency of the playback video signal. To achieve this control, the slow and search control signals from system control are applied to IC501-28, disabling the phase loop control. The system control also supplies a turn-on signal to the search correction IC at IC503-6, as shown in Fig. 7-16.

The horizontal sync (H-sync) signal from the video circuits (Chapter 12) is applied to IC503-15. A sample-and-hold circuit within IC503 develops a dc control voltage at IC503-22. This control voltage is added to the speed control signal from IC502-14 and the slow correction signal from IC504-22 (Chapter 8). The combined output is applied to the cylinder motor drive IC (Sec. 7-3). The added control voltage from IC503 shifts the speed of the cylinder motor slightly in search (both forward and reverse modes) to maintain the proper horizontal sync.

Cylinder servo search mode troubleshooting/repair. If the picture is out of horizontal sync only during search (either search forward or search reverse), look for the presence of the H-sync signal at IC503-15. If missing,

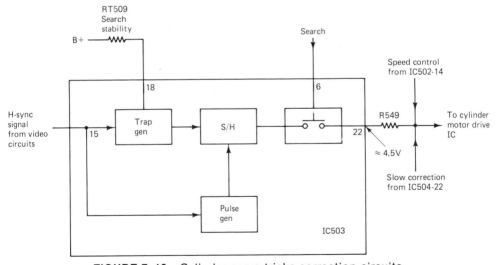

FIGURE 7-16 Cylinder servo tricks-correction circuits

check the sync separator circuitry in the video system (Chapter 12). If H-sync is present, look for a dc output of about 4.5 V at IC503-22. (Make sure that the search turn-on signal from system control is present at IC503-6.) Slowly rotate the search stability control RT509 while monitoring the voltage at IC503-22. If the voltage at IC503-22 does not change about ±1 V with adjustment of RT509, suspect IC503.

8

CAPSTAN SERVO

This chapter is devoted to troubleshooting and repair of VCR capstan servo circuits. The capstan servo must: (1) control the *speed* of rotation for the capstan (and thus the *tape speed*), (2) control the phase relationship (or speed vernier) of the tape and heads, and (3) provide drive for the capstan motor.

The overall functions of VCR servo systems are described in Sec. 1-4. The following sections describe details and characteristics of the capstan servo circuits.

8-1 BASIC CAPSTAN SERVO SYSTEM

Figure 8-1 is the overall block diagram of the capstan servo for a typical VCR (that shown in Fig. 1-1c). Note that the capstan servo is functionally similar, but not identical, to the drum/cylinder servo described in Chapter 7. Both servos use a form of PWM (pulse width modulation) to get the necessary speed and phase control. The use of PWM in servos is typical for most modern VCRs. (However, some VCRs use sample-and-hold circuits, also described in Chapter 7.) The servos of most modern VCRs also use the three-phase motor drives with Hall-effect control described in Chapter 7, so we do not repeat the motor drive circuits here. Instead, we concentrate on capstan speed and phase control functions that occur during the various playing times and operating modes.

As shown in Fig. 8-1, the phase-loop portion of the capstan servo is contained within IC501, and the speed-loop portion is within IC502. During

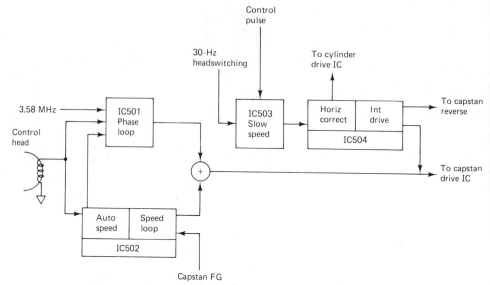

FIGURE 8-1 Overall block diagram of the capstan servo

playback, the *auto-speed select* circuits within IC502 monitor the control track signal (picked up by the control head) and determine the proper speed for the capstan motor. The correction voltage from the phase loop and speed loop are added together and passed to the capstan motor drive IC.

During slow motion and special effects, correction signals developed from IC503 and IC504 are passed to the capstan motor drive IC along with the speed-loop correction voltage. During slow-motion operation, the phase-loop portion of the servo (IC501) is inhibited, and phase correction is done by IC503.

8-2 CAPSTAN PHASE-LOOP OPERATION

Figure 8-2 shows the capstan phase-loop circuits for the VCR of Fig. 8-1. The capstan phase loop uses a 3.58-MHz signal input at IC501-25 as a reference signal for the PWM comparator system. During playback, the comparison signal is derived from the control track signal recorded on tape. During record, the comparison signal is derived from the capstan FG signal. This signal has three different frequencies, depending upon the playing time : 2160 kHz in SP; 1080 Hz in LP, and 720 Hz in SLP.

The control track signal and the capstan FG signal are passed through programmable divider IC503, which *does not* divide the signal in normal play or record modes, but does divide in the search modes. This division is necessary since capstan speed is increased, but phase lock must be maintained.

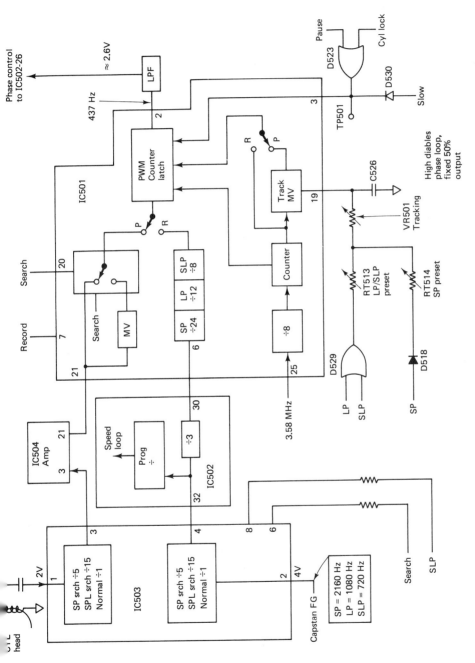

FIGURE 8-2 Capstan phase-loop circuits

To accommodate the increased frequency of the control track and capstan FG signals (SP is increased by 5; SLP by 15), the signals are divided down to normal frequencies so that both the phase and speed loops can control the capstan motor. The control track signal is passed from IC503-3 to IC504-3, where the signal is amplified and applied to the phase-loop IC501-21. The capstan FG signal at IC503-4 is passed through a divider within IC502 and applied to IC501-6.

During playback, the control track signal at IC501-21 is applied to the input of the PWM counter latch circuit. During record, the capstan FG signal is divided down to 30 Hz and applied to the PWM system. The capstan servo system adjusts capstan speed (and thus tape speed), thereby maintaining the divided-down capstan FG signal at 30 Hz.

To achieve proper *tracking,* the 3.58-MHz reference signal at IC501-25 is divided down and delayed by an MV. The time constant of the tracking MV is determined by the resistance and capacitance at IC501-19. The *front-panel tracking control* VR501 and preset controls RT514/RT513 determine the amount of delay to the reference signal in order to get proper phase of the control track signal, minimizing noise in the video signal.

The output of the PWM system is a 437-Hz signal appearing at IC501-2. *The duty cycle or pulse width of the PWM output signal is proportional to the phase error of the system.* The output is passed to an external low-pass filter circuit which removes the ac component and passes the dc component to the speed loop IC502 (Sec. 8–3). The dc component of the PWM signal is about 2.6 V.

During special effects, slow motion, pause, or cylinder-lock condition, a high is applied to IC501-3. This disables the PWM system and generates a fixed 50% duty cycle for the 437-Hz signal at IC501-2. In effect, the capstan phase control is removed from the circuit.

Capstan phase loop troubleshooting/repair. The most common symptom of problems in the capstan phase loop is a *noise bar* (or bars) floating through the picture, but with the picture properly synchronized. (If the capstan speed loop is malfunctioning, you will get excessive audio wow and flutter, along with picture instability, out of sync, etc.) If you suspect problems in the capstan phase loop, set the VCR to pause mode and check the 437-Hz PWM signal at IC501-2 for a 50% duty cycle. If absent or abnormal, suspect IC501.

If the PWM signal is good, return the VCR to play, and play back a known good tape or test tape. Check for the presence of the control track signal (CTL) at IC503-1. If the CTL signal is absent or abnormal, suspect a defective control track head, connector cable, or possible tape drive-path problem (tape not moving past the control head properly).

If the control track signal is good, check for a signal at IC503-3. If absent or abnormal, suspect IC503. Then check for a signal at IC504-21 and IC501-21. If the signal is absent or abnormal, suspect IC504.

If the signal at IC501-21 is good, check the 3.58-MHz reference signal at IC501-25.

If both the comparison signal (IC501-21) and reference signal (IC501-25) are normal, but the capstan phase loop appears to have no control of the capstan phase (noise bars, etc.) check that IC501-3 is low (normal phase loop operation). If IC501-3 is low, suspect IC501. If IC501-3 is high (in normal play), check the pause, slow, and cylinder-lock lines from system control (Chapter 14).

8-3 CAPSTAN SPEED-LOOP OPERATION

Figure 8-3 shows the capstan speed-loop circuits for the VCR of Fig. 8-1. The capstan speed loop develops a correction voltage resulting from a *difference in frequency* between the capstan FG signal and the divided-down 3.58-MHz signal. The capstan FG signal is passed through programmable divider IC503 to IC502-32. IC503 divides the capstan signal by 5 in the SP search mode and by 15 in the SLP search mode. The capstan FG signal is not divided by IC503

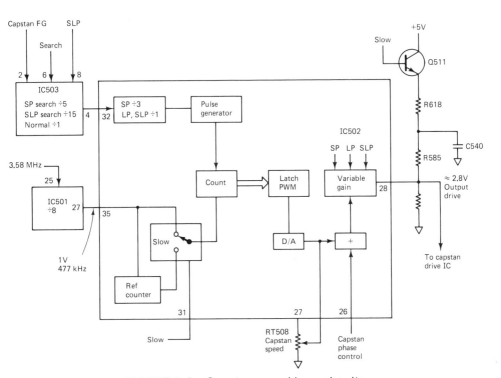

FIGURE 8-3 Capstan speed-loop circuits

in normal playback. The FG signal is further divided within IC502 and passed to a *digital counter,* which is part of the PWM system of IC502.

The 3.58-MHz reference signal is first divided by 8 in IC501 and passed to IC502-35 as a 477-kHz signal. The divided-down signal is then processed and applied to a counter within IC502 for comparison with the divided-down capstan FG signal. The PWM circuit develops a dc output voltage corresponding to the difference in frequency between the capstan FG signal and the 3.58-MHz reference.

Capstan speed control RT508 at IC502-27 adjusts the level of the dc correction voltage for the speed loop. The correction voltage is combined with the capstan phase control signal (Sec. 8-2) by an adder within IC502. The combined speed and phase-loop voltages are applied to a variable amplifier. The gain of the amplifier is determined by the auto-speed select circuit within IC502. The output at IC502-28 is passed to the capstan motor drive IC.

During slow motion, the capstan servo speed loop is disabled by a high at IC502-31. The output drive voltage is also pulled up to about 3.5 V by turning on Q511. In still operation, the output drive voltage is pulled to ground by IC504, as discussed in Sec. 8-4.

Capstan speed-loop troubleshooting/repair. The most common symptoms of problems in the capstan speed loop are excessive audio wow and flutter, or picture instability (out of sync), or both. The first step in capstan speed-loop troubleshooting is to disable the phase-loop circuits by connecting TP501 (IC501-3, Fig. 8-2) to B+. This disables the phase-loop operation and sets the output of the PWM system to 50%. Check for about 2.6-V input to the capstan speed loop at IC502-26. If absent or abnormal, suspect a problem with the 50% duty cycle square-wave output from the phase loop (Sec. 8-2) or in the low-pass filter between the phase loop and speed loop.

If the input from the phase loop at IC502-26 is normal, then monitor the capstan free-run waveform while slowly rotating capstan speed control RT508 (Fig. 8-3). Adjust RT508 to get the proper sampling pulse and trapezoid lockup. (The service literature describes the exact procedure. Follow the procedure! The waveforms involved are similar, but not identical, to those shown in Fig. 7-3.)

If the capstan speed loop *can be* locked in (picture stable, minimum, or no wow or flutter) by adjustment, with the phase loop disabled, suspect the phase-loop components (Sec. 8-2). If you cannot correct capstan speed-loop problems by adjustment of R508, look for the correct inputs to the capstan speed loop at IC502-32 and IC502-35. If the signals appear normal, suspect IC502. If the signals are absent or abnormal, look for a 3.58-MHz reference at IC501-25 and for a capstan FG signal at IC503-2. If either signal is absent or abnormal, trace the 3.58-MHz reference line and/or capstan FG line. If the inputs to IC501 and IC503 are good, suspect IC501 and IC503.

8-4 CAPSTAN SLOW/STILL OPERATION

Special circuits are often used in modern VCRs during slow/still operation to control the capstan servo. The following paragraphs describe such circuits for both slow/still start and slow/still drive.

8-4.1 Capstan slow/still start operation

Figure 8-4 shows the capstan servo operation during slow/still start. Figure 8-5 shows the related waveforms.

When the slow command is first applied (from the front panel or remote unit), the VCR system control applies a high to IC503-10. When the next positive-going control track pulse appears at IC503-1, a high output (slow enable) is generated at IC503-7. This high is applied to the capstan phase-loop circuit through D530, turning off the phase-loop IC as discussed in Sec. 8-2. The high is also applied to IC504-8 to turn on the slow-motion processing circuits (Sec. 8-4.2). The slow enable signal is also applied to the slow operation pulse circuit within IC503, as shown in Fig. 8-4.

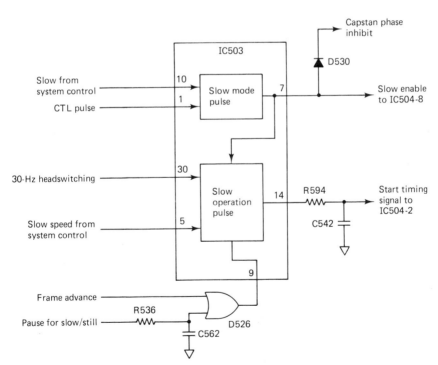

FIGURE 8-4 Capstan slow/still start circuits

192   Capstan Servo

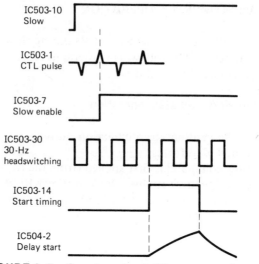

FIGURE 8-5   Capstan slow/still start waveforms

When the high (slow enable) occurs, the slow operation pulse circuit counts two 30-Hz head-switching pulses and generates a high at IC503-14. This high is called the *start-timing signal,* and determines the beginning of slow-down operation of the capstan motor. The start-timing pulse is a fixed duty-cycle output of about two 30-Hz head-switching periods (Fig. 8-5). The start-timing pulse is integrated by R594/C542 and is applied to IC504-2.

The repetition rate of the delayed start waveform is the function of an up/down counter within IC503, and is controlled by the slow-speed logic signal at IC503-5 from system control. When slow-speed information is passed to IC503-5, the up/down counter determines how many 30-Hz head-switching pulses occur before the start-timing pulse is generated at IC503-14.

During still-frame operation, the up/down counter is inoperative, and a one-shot MV is activated by the frame-advance signal applied to IC503-9 through D526. The MV generates a pulse at IC503-14 to move the tape to the next frame, as discussed next.

8-4.2   Capstan slow/still drive operation

Figure 8-6 shows the capstan servo operation during slow/still drive. Figure 8-7 shows the related waveforms.

As discussed in Sec. 8-4.1, IC503 generates a slow enable signal that is applied to IC504-8. When the next control track pulse occurs at IC504-3, the slow-tracking MV is turned on. The time delay of the tracking MV is determined by the resistance of the slow-tracking control at IC504-4. The tracking

Capstan Slow/Still Operation   193

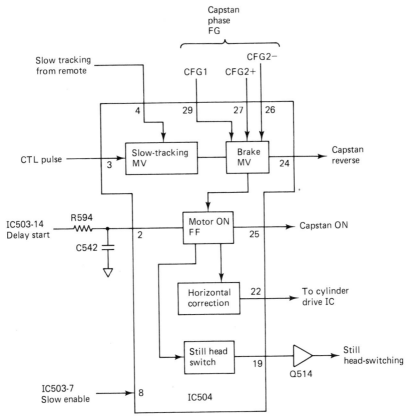

FIGURE 8-6 Capstan servo operation during slow/still drive

MV turns on the brake MV, which applies a capstan-reverse signal pulse to the capstan motor drive IC. This causes the capstan motor to slow down immediately and then stop. The width of the brake pulse is determined by IC504 monitoring the positive and negative capstan phase FG signal at pins 26, 27, and 29 of IC504. By detecting the order in which test pulses occur, IC504 knows when the capstan motor has reversed direction. At that instant the brake pulse is terminated, preventing the capstan motor from rotating in reverse.

The brake pulse is also applied to the motor-on FF within IC504. This FF develops a positive-going drive pulse that is applied as the capstan-on input to the capstan motor drive IC through IC504-25. During the time that IC504-25 is high, the capstan motor drive IC applies power to the capstan motor.

The motor-on FF also generates the capstan-on pulse as a result of the delayed-start signal applied to R594/C542 at IC504-2 (as discussed in Sec. 8-4.1). The delayed-start signal starts the capstan motor rotation from the still mode.

194 Capstan Servo

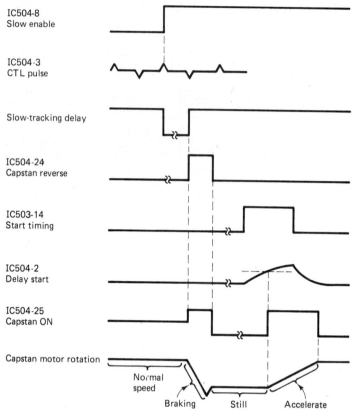

FIGURE 8-7 Capstan servo waveforms during slow/still drive

Slow motion. Slow motion is accomplished by moving the capstan motor (and tape) through four phases *between* control pulses. The four phases are: *normal playback speed, braking, still,* and then *acceleration* back to normal speed.

The capstan motor operates at normal playback speed when the capstan-on pulse is high at IC504-25. When a control pulse appears, and the brake pulse is applied at IC504-24 (at a time determined by the slow tracking control), the capstan motor slows down but does not go into reverse. Instead, the capstan motor moves forward slightly before going into still (when IC504-25 goes low).

At a time determined by the slow-tracking MV (which, in turn, is determined by the slow-tracking control) IC504-25 again goes high, accelerating the capstan motor back to normal playback speed. The motor remains at normal speed until the next control track pulse is detected (at IC504-3), which generates a brake pulse. This sequence of normal speed, braking, still, and then acceleration repeats as long as the slow-enable signal is applied to IC504-8.

Still operation. In still (also called *still field* or *freeze frame* in some VCRs), the capstan motor is decelerated to a stop (with a braking signal) as in slow motion. However, the up/down counter within IC503 is made inoperative, and the one-shot MV is activated by the frame-advance signal at IC503-9 (Fig. 8-4). So the capstan motor remains stopped and does not go into acceleration until the frame-advance signals are applied by the user (from the front-panel or remote-control frame-advance button).

During still operation, and during the still portion of slow motion (immediately after braking), a positive pulse is generated from IC504-19 (Fig. 8-6). The pulse is buffered by Q514 and applied to the luminance/chroma video circuits. This allows the fifth head to be switched into the circuit during still, as described in Sec. 6-4.

Capstan slow/still drive troubleshooting/repair. If operation of the VCR is good in all modes, but absent or abnormal in slow/still, first check that the slow-enable output pulse at IC503-7 (Fig. 8-4) is present when the slow button (VCR front panel or remote) is pressed. If missing, check the input signals at pins 1 and 10 of IC503. If missing, trace the input signals back to the control head and system control. If both input signals are present, suspect IC503.

Next check for the start-timing pulse at IC503-14. If missing, check the input signals at pins 5 and 30 of IC503. If missing, trace the input signals back to the head-switching and system control circuits. If both input signals are present, suspect IC503.

Next check for the presence of a control track input pulse at IC504-3 (Fig. 8-6). If missing, trace the signal back to the control head. Then check for delayed-start and slow-enable inputs at pins 2 and 8 of IC504, respectively. If these inputs are missing, suspect IC503. If all three input signals are present, check for a capstan-reverse (brake) pulse output at IC504-24. If missing, suspect IC504. If present, then check for a motor-on signal at IC504-25 occurring at the same time as the capstan-reverse braking pulse. If missing, suspect IC504.

If both the brake pulse and capstan-on pulse are present and occur at the same time, as shown in Fig. 8-7, the problem is most likely in the capstan motor drive IC or the capstan motor.

If slow motion appears to be good but there is no frame advance, check for frame-advance input signals to OR-gate D526 (Fig. 8-4) when the frame-advance button is pressed. Also check for frame-advance input pulses at IC503-9. If these pulses are absent or abnormal, track back from D526 to the frame-advance buttons through system control.

196 Capstan Servo

If there is slow motion, but the slow-motion tracking control has little or no effect (you cannot get noise bars out during slow motion), trace back from IC504-4 (Fig. 8-6) to the slow-motion tracking control.

8-5 USING FRAME ADVANCE AND THE CAPSTAN SERVO IN STILL

Some VCRs use the frame-advance signal and the capstan servo to get proper positioning of the video head on the tape track during still or freeze-frame mode. Such is the case with the four-head VCR described in Sec. 6-2. The circuit involved in this operation is shown in Fig. 8-8. The related pulse timing is shown in Fig. 8-9.

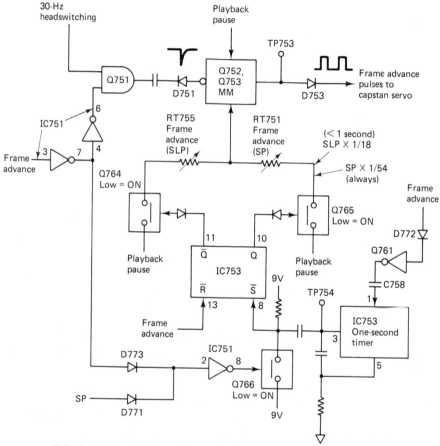

FIGURE 8-8 Frame-advance pulse generation circuits (during still)

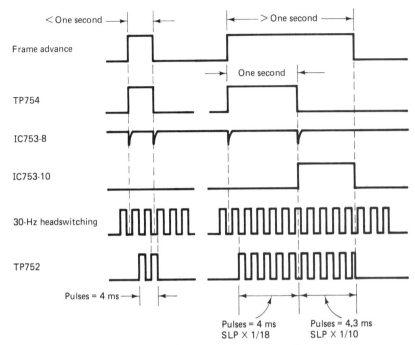

FIGURE 8-9 Frame-advance pulse generation waveforms (during still)

One of the problems with early-model VCRs is that noise bars often appear in the picture during still operation. This indicates improper video-head-to-tape-path positioning. Proper positioning can be achieved by applying a frame-advance pulse to the capstan servo, allowing *partial rotation* of the capstan motor. This advances the tape by a very small amount and positions the special-effects video heads properly over the tape tracks.

In the circuit of Fig. 8-8, the width of the frame-advance pulse is set by the time constant of the Q752/Q753 MM (monostable multivibrator). This MM is triggered from the falling edge of the 30-Hz head-switching signal. The 30-Hz signal is applied to the MM through AND-gate Q751 and D751. AND-gate Q751 is turned on by the frame-advance signal from system control through the inverters of IC751, so the frame-advance pulse is available (for still operation) only when the frame-advance button is held down.

When the frame-advance button is held continually, the MM generates a series of frame-advance pulses at a 30-Hz rate to the capstan servo system, rotating the capstan motor and moving the tape. The width of the frame-advance pulse is set by the time constant applied to the MM input through frame-advance controls RT755 and RT751.

During still operation at the SP playing speed, the frame-advance speed is preset to a speed 1/54th of the normal SP tape speed. During SLP still operation, the frame-advance pulse occurs at two different pulse widths, as shown in Fig. 8-9. For periods when the frame-advance button is held down for less than one second, the rate for SLP frame-advance pulse is 1/18th normal SLP tape speed. For periods when the frame advance is held longer than one second, the SLP frame advance speed is set to 1/10th of the normal SLP speed.

During still operation, either RT751 or RT755 is connected to a high from the playback pause line through corresponding switches Q764 or Q765. This connects either RT751 or RT755 to a B+ source (playback pause line) and permits the corresponding frame-advance control to set the MM time constant (which, in turn, sets the amount of capstan/tape movement).

In normal SP playback, system control applies a fixed-level high through D771 to IC751-2. The high is inverted to a low by IC751 and is applied to Q766, turning on Q766 and applying 9 V to the active-low input of IC753 at pin 8. The frame-advance pulse, which is normally low at this time, is applied to IC753-13 (the active-low reset input). When IC753-8 goes high (from the 9 V), the output at IC753-11 is high and the output at IC753-10 is low, no matter what level is applied to IC753-11 (by frame advance). IC753 remains fixed in this state as long as the SP high is applied to Q766 through D771 and inverter IC751. This connects SP control RT751 to B+ and sets the frame-advance tape speed to 1/54th of the normal SP speed.

In SLP, when the frame-advance button is pressed, a high is applied that triggers a one-second timer within IC753. The frame-advance pulse is also applied to IC751-3, where the pulse is inverted and appears as a low at IC751-7. This low is applied through D773 and inverted by IC751 to a high. The high turns off Q776 (removing the fixed 9 V used in SP) and connects IC753-8 to the output of the one-second timer, so that IC753 is controlled by the timer.

As shown in Figs. 8-8 and 8-9, the frame-advance high turns on Q751 to *start triggering* the frame-advance MM Q752/Q753. The pulse width of the MM output (and thus the amount of capstan/tape movement) is determined by the state of FF IC753. Initially, IC753-10 is low (because of the high at IC753-13), turning on Q765 and connecting RT571 to B+ (as in the SP mode). This sets the frame-advance tape speed to 1/18 of the normal SLP speed. If the frame-advance button is held in longer than one second during SLP, the output of one-second timer at IC753-3 goes low, generating a negative-going pulse to pull the set input at IC753-8 low. This causes the output at pins 10 and 11 of IC753 to toggle, turning on Q764 and turning off Q765. This connects RT755 to B+, setting the frame-advance tape speed to 1/10th of the normal SLP speed. (An increase in the frame-advance pulse increases capstan/tape movement.)

Capstan servo still operation troubleshooting/repair. The repetition rate of the frame-advance pulses in the circuit of Fig. 8-8 are very difficult to monitor with a conventional oscilloscope. Generally, you need a *logic probe* as discussed in Chapter 2. The probe makes it relatively easy to detect the presence or absence of the pulses (which is generally all you need to trace through the microcomputer and/or system control sections of most VCRs).

In the circuit of Fig. 8-8, start by checking for a frame-advance enable signal to AND-gate Q751. If this signal is missing, with the frame-advance button pressed, suspect one or more of the inverters within IC751. If the signal is present but there is no signal applied to MM Q753/Q754, check the 30-Hz head-switching signal, and the signal path between Q751 and MM Q752/Q753 (including D751). If there is a trigger to the MM, check for the presence of the MM output at TP753. If missing, check for proper operation of switch transistors Q754 and Q755, as follows.

During still operation and frame advance, one of transistors Q754 and Q755 is active, applying B+ to the appropriate control RT751 or RT755. If both Q754 and Q755 are off, check the outputs of IC753 (pins 10 and 11). One output should always be high, while the opposite output is low (pin 10 high and pin 11 low, or vice versa). If not (both outputs high, or both low), suspect IC753. If one output is high and the other low, but there is no change in outputs when switching between SP and SLP, or when pressing the frame-advance button, check for proper inputs at pins 8 and 13 of IC753. Trace the lines back to their sources (through system control to the operating controls). If the outputs from IC753 and pins 10 and 11, are correct, but Q764 or Q765 are not on, suspect Q764/Q765.

During SLP frame-advance operation, if an increase in frame-advance pulse generation is not observed after one second, check if Q766 is turned off and if a pulse appears at TP754. If Q766 is not turned off, check for frame-advance pulses through D773 and two inverters in IC751. If missing, trace the pulses back to their source. Also check that an SP pulse is not applied to lock Q766 on. (This should occur during SP, but not SLP.) If Q766 is off, but there is no pulse at TP754, suspect IC753 or the associated capacitor and resistor at pins 3 and 5 of IC753.

8-6 OPERATING THE CAPSTAN SERVO AT DOUBLE SPEED

In some VCRs the capstan servo can be operated at double speed (or at some much-higher speed) during certain modes of operation. Such is the case of the four-head VCR described in Sec. 6-2. The circuit involved in this operation is shown in Fig. 8-10. The VCR is capable of *double speed* in SP or SLP (via a wired remote unit in this case).

To get double-speed operation, the capstan motor speed is increased by a factor of 2. In normal operation, the capstan speed is controlled by the value of capstan speed control RT608 connected at IC604-1. During double-speed operation, an additional resistance is connected in parallel with RT608.

200 Capstan Servo

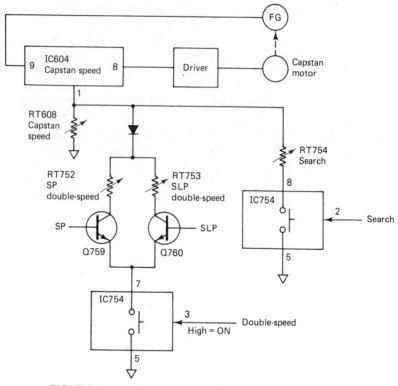

FIGURE 8-10 Capstan servo double-speed circuits

Depending on the selected playback speed or playing time (SP or SLP), either Q759 or Q760 turns on due to a high at the corresponding base. The emitters of Q759 and Q760 are connected to ground by a switch within IC754 when a double-speed command is applied to IC754-3. In SP, RT752 is connected in parallel with RT608 when Q759 is turned. In SPL, RT753 is connected across RT608 when Q760 is turned on. This changes the resistance at IC604-1 and doubles the speed of the capstan and tape.

Also note that the capstan speed is increased during *search* operation by connecting a control RT754 in parallel with RT608. This is done by another switch within IC754 when a search command is applied at IC754-2.

Capstan servo double-speed troubleshooting/repair. If the capstan motor operates properly in normal-speed modes, but not in double-speed or search modes, the problem is most likely in the switching transistors Q759/Q760 and/or the switches within IC754. If the problem is in search mode only, check for a high at IC754-2. If the high is missing, trace back to system control. If the high is present, check that pin 8 and RT754 are connected to ground through IC754.

Operating the Capstan Servo at Double Speed   201

If the problem is in double-speed mode, check for a high at IC754-3, and that the emitters of Q759/Q760 are connected to ground. If the high is missing, trace back to system control. If the high is present, suspect IC754. If the emitters are properly grounded, check for a low at one of the transistor outputs. Since Q759/Q760 are NPN, one of the collectors should go low and connect the corresponding control to ground (or near ground) when a high is applied at the base. If the high is missing, trace back to system control from the corresponding base. If the high is present at the base but the collector does not go low, suspect the corresponding transistor.

Note that if double-speed or search-speed operation appears to be good (tape moving at a much higher speed), but you suspect that the speed is not correct (and this is difficult to judge), you must check adjustment of the controls as described in the service literature. However, do not go in and readjust the speed controls (or any other controls on the VCR) on suspicion alone, or without using the service literature procedures. This is often the technician's version of "operator trouble"!

8-6.1 Double-speed tracking problems

In some VCRs the tracking control time constant is changed in certain modes. For example, in the VCR described in Sec. 6-2, the tracking time constant is changed slightly to compensate for tape stretching in the SP double-speed playback mode. The circuits involved are shown in Fig. 8-11.

Operation of the tracking circuit is obtained by controlling the amount of reference signal delay processed within the capstan phase loop IC602. The time constant at IC602-18 is controlled by variable resistor RV901 (the front-panel tracking control) and the normal tracking preset control RT601, which is connected to B+ through Q601.

During normal-speed SP operation, the double-speed command from system control to Q763 is low, producing a low at the output of Q763 and a low at the input of Q601. This low turns on Q601, connecting RT601 to RV901. The low from Q763 is also inverted to a high by IC752 and applied to Q602. This high keeps Q602 off and keeps double-speed tracking control RT602 out of the circuit.

During double-speed SP operation, the double-speed command from system control to Q763 is high, producing a high at the output of Q763, a high at the input of Q601, and a low at the input of Q602. This turns off Q601 and turns on Q602, disconnecting RT601 and connecting RT602 into the circuit. With RT602 in series with RV901, the tracking control time constant is changed slightly to compensate for tape stretch in the SP double-speed mode.

Note that AND-gate Q763 is controlled by the SP command from system control. In SP, the command is high. This high is inverted to a low by IC754 and applied to the inverting input of Q763. This turns on Q763 and permits the double-speed command to pass to the output of Q763. When not in SP,

202 Capstan Servo

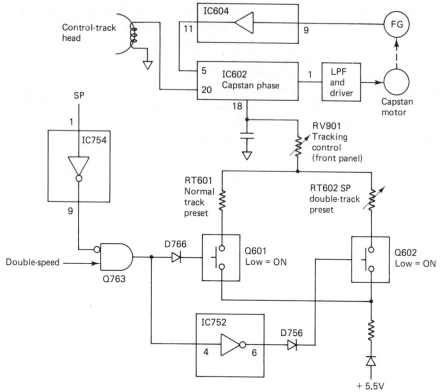

FIGURE 8-11 Double-speed tracking circuits

the SP command line goes low, disabling Q763 and preventing the double-speed command from passing to Q601/Q602. In any playing speed but SP, the output of Q763 remains low, turning on Q601 and turning off Q602, no matter what the state of the double-speed command.

Double-speed tracking troubleshooting/repair. If tracking appears to operate properly at normal speed in SP, but not at double speed, the problem is most likely in the switching transistors Q601/Q602, the inverters IC752/IC754, or the double-speed AND gate Q763. Check that both the SP and double-speed commands from system control are present, and that IC754 inverts the SP commands to a low at the inverting input of Q763.

If the inputs to Q763 are absent or abnormal, trace back to the source. If the inputs are good, but the output of Q763 remains low, suspect Q753.

If the output from Q763 is high, check that the input to Q602 is low. If not, suspect IC752. If the input to Q602 is low, but RT602 is not connected into the circuit, suspect Q602.

9

SPECIAL EFFECTS

This chapter is devoted to troubleshooting and repair of VCR special-effect circuits. There is no standardization on special effects for modern VCRs. However, as a general rule, the more modern the VCR, the more special effects are available. (The same is true for the higher-priced VCRs!) In this chapter we describe the special-effect circuits for a typical modern VCR (that shown in Fig. 1-1a). This represents a cross section of special-effect circuits, since the VCR is top of the line.

9-1 SPEED SEARCH

Figure 9-1 shows the speed-search circuit. (Note that speed search is called *fast forward* and *fast reverse,* or *fast-forward search* and *fast-reverse search* in some VCRs.) No matter what the mode is called, speed search allows the tape to be transported past the heads at a high rate of speed, while maintaining head output and a *reproduced picture*. Audio, which is unintelligible at such high speeds (Martian voices), is muted during speed search.

During speed search, the *relative* tape-to-head angle is altered. This generates *noise bars* moving through the picture. The bars are proportional in number to tape speed. While noise bars occur in speed-search mode for most VCRs, they can be reduced in intensity and made relatively stable, making them less objectionable. The circuit of Fig. 9-1 does both of these functions, using the 4X video head system (Sec. 6-3) in conjunction with the capstan servo control (Chapter 8), during speed search.

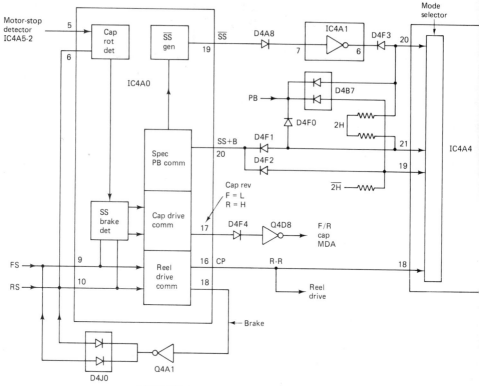

FIGURE 9-1 Speed-search circuits

As discussed in Chapter 8, the capstan servo is designed to operate with specific FG and CTL input frequencies, both of which are proportional to capstan/tape speed. Since capstan and tape speeds are increased in speed search, a means is provided to return the FG and CTL frequencies back to their original values and thus maintain proper capstan servo control. This is done by *dividing* the increased frequencies of both signals by a *factor equal to the increase* in capstan/tape speed.

For example, in 2H, capstan/tape speed is increased to seven times (7X) normal, requiring a division of the FG and CTL frequencies by a factor of 7, in order to return both to their original value. In 6H, capstan/tape speed is increased to 9X normal, requiring a division of the FG and CTL frequencies by a factor of 9.

Division of the FG and CTL frequencies is done by the FG/CTL divider. This divider is limited to division by 7 in 2H and division by 9 in 6H. The division factor is determined by logic within the *automatic speed selector* portion of the capstan servo IC.

The capstan motor speed is increased during either forward or reverse speed search. This is done by dividing the original FG and CTL frequencies so that the FG and CTL signals *appear* to be at a much lower frequency than normal. The capstan servo reacts to this apparent change in speed by increasing capstan/tape speed until the FG and CTL frequencies are normal. However, this condition is not met until the capstan/tape speed is 7X(2H) or 9X(6H) the normal playback speed.

The capstan motor is driven in a forward or reverse direction, depending on the speed-search mode selected. *Electromagnetic braking* (called *electronic braking* in some VCR literature) is applied at the termination of speed search. This is done by applying a momentary reverse drive to the capstan motor at the end of forward search (and forward drive at the end of reverse search).

The *reel motor* drive (Chapter 5) must also be considered during speed search. For example, in forward search (FS), the TU motor must be driven, while in RS, the SU motor is operated. The logic that controls the direction of rotation for both the capstan motor and reel motors is developed by the speed-search circuit.

The drum/cylinder servo (Chapter 7) is also involved in speed search. The increase in capstan motor speed affects the relative tape-to-head speed. If the drum/cylinder servo continues to operate at 30 rps (1800 Hz), the relative tape-to-head speed is changed considerably if the capstan/tape speed changes. For example, during a 6H FS mode, the relative tape-to-head speed is reduced (since tape movement and video head rotation are in the *same* direction). This has the undesirable effect of decreasing the horizontal sync frequency by 6%. Reverse search has the opposite effect, since tape movement and head rotation are in opposing directions. This increases the relative tape-to-head speed, increasing the horizontal sync frequency by 6%.

Either condition can cause a loss of horizontal sync on certain monitoring devices when speed search is selected. The appropriate tape-to-head speed can be recovered by increasing the drum/cylinder motor speed in FS and decreasing motor speed in RS.

9-1.1 FG/CTL divider and mode selector

Division of the FG and CTL signals during speed search is done by circuits within the capstan servo IC4A4, shown in Fig. 9-1. The division ratios are determined by the mode selector portion of IC4A4 upon receiving commands from the special-effects IC4A0. Speed-search commands are output from pins 16-20 of IC4A0 and are applied to the mode selector at pins 18-21 of IC4A4.

Since speed search is used only in playback, the mode selector inputs at pins 19-21 of IC4A4 are held low during any other operational mode by a low PB command applied to D4B7 and D4F0, effectively disabling the speed-search function. (The PB line is high in playback only.)

During a valid speed-search mode (PB high), the FG/CTL division ratio (7 or 9) is determined by the mode selector input at pins 19-21 of IC4A4. In

2H, pin 19 is held low by the low on the $\overline{2H}$ line (which is low in 2H), establishing a division factor of 7. In 6H, pins 20 and 21 of IC4A4 are held low by the low on the 2H line (which is high in 2H), establishing a division factor of 9.

The CP R-R (capstan reel-reverse) input at pin 18 of IC4A4 informs the mode selector of the rotational direction for the capstan motor (and thus the direction of tape travel through the transport). CP R-R is low during FS and high during RS, telling the mode selector to increase drum/cylinder speed for FS or decrease it for RS.

9-1.2 Forward search (FS)

When forward search is selected from normal playback, system control produces a high on the FS line at IC4A0-9. This turns on a special playback command circuit within IC4A0, which responds by generating a high at the pin 20 output (SS+B) and a low at pin 19 ($\overline{SS}$) of IC4A0.

The capstan direction command at pin 17 of IC4A0 goes low, disabling D4A4 and Q4D8, placing a high on the F/R (forward/reverse) line to establish forward capstan/tape movement. The capstan motor rotation (CP R-R) output at pin 16 is also low and, following application to the reel drive system, establishes and maintains drive to the TU reel motor for the duration of the FS mode. the CP R-R command also applies a low to the mode selector at pin 18 of IC4A4, designating that the drum/cylinder motor speed be increased.

In 2H, a high 2H command is applied to pins 20 and 21 of IC4A4. Since the $\overline{SS}$ output is low, D4A8 is disabled. Since SS+B is high, D4F1 and D4F2 are also disabled. The input to pin 19 of IC4A4 is held low by the low on the $\overline{2H}$. With highs at pins 20 and 21 and a low at pin 19, IC4A4 produces the necessary commands to establish a division ratio of 7 in the FG/CTL divider.

In 6H, the 2H line goes low, holding the inputs at pins 20 and 21 of IC4A4 low. The $\overline{2H}$ line is high, placing a high at pin 19 of IC4A4. The low at pins 20 and 21, in conjunction with the high at pin 19, causes IC4A4 to produce a division ratio of 9 in the FG/CTL divider.

The truth table shown in Fig. 9-2 shows the IC4A0 output states for various speed search modes. As discussed in Sec. 9-1.6, such a truth table is very convenient for troubleshooting problems in the speed-search circuits.

9-1.3 Forward search (FS) brake

When the FS button is released at the completion of forward search, a braking force is applied to the capstan motor by momentarily reversing the direction of rotation. This is done by placing a high on the CAP-REV (capstan-reverse) line at pin 17 of IC4A0. At the release of the FS button, the high-to-low transition at the FS input IC4A0-9 triggers the SS brake detector circuit and the

Mode \ Pin	Stop det 5	RS 6	FS 9	RS 10	CP R-R 16	Cap rev 17	Brake 18	$\overline{SS}$ 19	SS+B 20
FS	L	L	H	L	L	L	L	L	H
FS brake	H*	L	L	L	L	H	H	H	L
RS	L	H	L	H	H	H	L	L	H
RS brake	H*	L	L	L	L	L	H	H	L

*Momentary high as the capstan motor stops

FIGURE 9-2 Speed-search truth table

capstan drive command circuit within IC4A0. This produces a high on the CAP-REV line, turning on D4A4. The high is inverted to a low by Q4D8, placing a low on the F/R line, reversing drive to the capstan motor, and bringing motor rotation to a halt. During the braking process, a high is also placed on the brake line at pin 18 of IC4A0. This high is inverted to a low by Q4A1. The low is applied to both the FS and RF inputs, pins 9 and 10 of IC4A0, for the duration of the braking period.

Once the capstan motor stops, the braking force is removed immediately to prevent overshoot. This is done by the motor stop detector within IC4A5, which detects the point at which the motor stops and then generates a high at IC4A0-5 (the capstan rotation detector). The output at IC4A0-17 drops to a low, restoring forward rotation to the capstan motor. The truth table for FS braking is shown in Fig. 9-2.

9-1.4 Reverse search (RS)

During RS, a high is applied to pin 10 of IC4A0 by system control. This produces a high on the SS+B line, pin 20, and a low at the $\overline{SS}$ line, pin 19, of IC4A0. Logic determining the division ratio of the FG/CTL divider is applied to the mode selector within IC4A4, as described for forward search (Sec. 9-1.2).

The high RS at pin 10 of IC4A0 is also applied to pin 6 of IC4A0. This causes a high at IC4A0-17, the CAP-REV line. The high is inverted to a low by Q4D8, placing a low on the F/R line, causing the capstan motor to reverse. As the capstan motor begins to rotate in reverse, the CP R-R line at pin 16 of IC4A0 goes high, informing the reel drive system of reverse rotation. This causes the SU motor to rotate in reverse and take up the tape. The CP R-R logic is also applied to the mode selector within IC4A4, reducing speed of the drum/cylinder motor.

208   Special Effects

9-1.5   Reverse search (RS) brake

When the RS button is released at the completion of reverse search, rotation of the capstan motor is momentarily halted, forward rotation is restored, and normal servo control is resumed. Release of the RS button places a low at pins 6 and 10 of IC4A0. IC4A0 responds to the high-to-low transition by generating a high on the $\overline{SS}$ line, pin 19, a low on the SS+B line, pin 20, and a high on the brake line, pin 18, of IC4A0.

The low at pin 6, in conjunction with the low at pin 10, places a low on the CAP-REV line, pin 17, of IC4A0. This reverses drive to the capstan motor, bringing motor rotation to a stop. At that point, the motor stop detector in IC4A5 again places a momentary high at pin 5 of IC4A0, terminating the RS braking action and restoring normal servo control.

9-1.6   Speed search troubleshooting/repair

Troubleshooting the speed-search circuits shown in Fig. 9-1 is mostly a matter of checking inputs and outputs using an oscilloscope or logic probe. The truth table of Fig. 9-2 can be most helpful in this process. For example, in FS, all pins of IC4A0, except 9 and 20, should be low. If the inputs (pins 9 and 10) are correct, but the outputs are not, suspect IC4A0. Of course, if the inputs to IC4A0 are not correct, the problem is likely in system control.

If the outputs from IC4A0 are correct, but inputs to IC4A4 are not, check the diodes and inverters between the two ICs. Also check the system control inputs applied to IC4A4, such as the PB, 2H, and $\overline{2H}$. If inputs to IC4A4 are correct, but the response is not (no forward search, no reverse search, etc.) suspect IC4A4. Of course, all of this is based on the assumption that the servos (capstan, drum/cylinder, reel) operate properly in modes other than speed search. If not, check out the servos before you go into the speed-search circuits!

If there is a problem in braking, check for momentary highs at pin 5 of IC4A0 when the FS and/or RS buttons are released. If not, suspect the motor stop detector circuits within IC4A5.

9-2   SLOW MOTION (CONTINUOUSLY VARIABLE)

Figure 9-3 shows the basic slow-motion circuits in block form. Figure 9-4 shows the slow-motion circuit details. As in the case of the VCRs discussed in Chapters 6 through 8, slow motion for the circuits described here is made possible by a sequence of capstan motor *run-coast-brake-still* cycles. In the Fig. 9-3/9-4 circuits, the sequence is repeated continuously when slow motion is selected by a remote-control unit. One unique feature of the circuits discussed here is that slow motion is *continuously variable* with speeds ranging

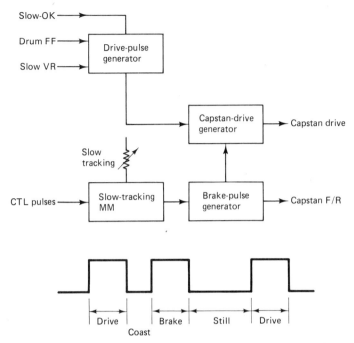

FIGURE 9-3 Basic slow-motion (continuously variable) block diagram

from about 1/7th to 1/30th of normal playback speed. (The slow-motion rate is fixed in many modern VCRs.) The slow motion is considered noiseless at both 2H and 6H tape speeds. If there is any residual picture noise, the *slow tracking* control can be adjusted to remove noise (just as the tracking control is used for normal playback speeds).

As shown in Fig. 9-3, the slow-motion circuit generates a series of run-coast-brake-still cycles at a repetition rate determined by the slow-motion speed control settings on the remote unit. The run, coast, and brake segments of the slow-motion cycles are at a fixed duration for a given playback speed. The repetition rate for a particular series of complete slow-motion cycles is determined by varying the *duration of the still segment*. Short-duration still segments increase the repetition rate of the slow-motion cycles, and vice versa. Additional service adjustments preset the *duration of the run* segment at a specified value to coincide with playback speed.

When slow motion is selected, a high SL-OK (slow OK) command turns on the drive pulse generator which, in turn, triggers on the positive rise of the next recurring drum FF pulse at the input. The output of the drive pulse generator goes high and remains high for a period of time determined by the setting of the slow VR. The output then automatically drops to a low, and one drive pulse is completed.

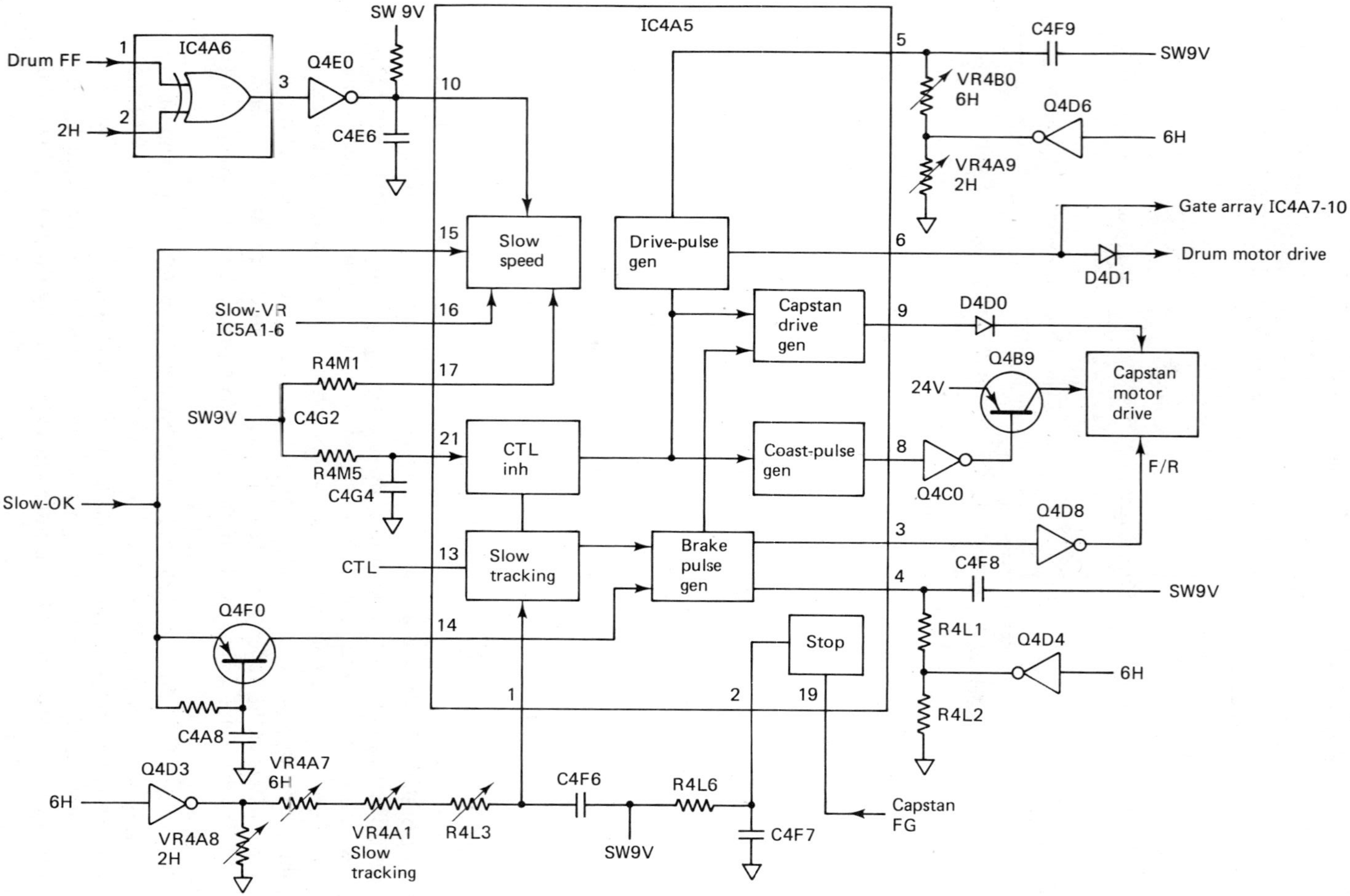

FIGURE 9-4 Slow-motion (continuously variable) circuits

Slow Motion (Continuously Variable)

The duration of the drive pulse, when applied to the capstan drive generator, establishes the active capstan run time during each slow-motion cycle. Drive is terminated at the fall of the drive pulse, and the coast segment begins. During coast, the capstan motor (without power) continues to rotate due to the inertia provided by rotation of the capstan flywheel. Synchronization with the CTL pulse occurs during coast. As the capstan motor slows, the next recurring CTL pulse triggers a slow tracking monostable multivibrator, the on-time of which is determined by the setting of the slow tracking control.

Coast time continues until the slow-tracking MM returns to an off-state, triggering the brake pulse generator and developing a high braking pulse. This braking pulse is inverted to a low and applied to the capstan F/R line, driving the capstan motor in reverse direction to provide instantaneous braking. The braking pulse is terminated at the capstan motor stop point, and the still segment of the slow-motion cycle is initiated. The still segment is maintained for a period of time determined by the slow-motion speed control. At the end of each still segment, the slow-motion cycle is repeated.

9-2.1 Run segment

When slow motion is selected, rotation of the capstan motor must be halted *before the run segment* of the slow-motion cycle can begin. When slow motion is first selected, the SL-OK line goes high and is applied to Q4F0. Although the emitter of Q4F0 goes high immediately, the base remains low until C4A8 charges, placing the base of Q4F0 at a lower potential than the emitter. As a result, Q4F0 conducts momentarily, placing a momentary high at the still input, pin 14 of IC4A5. This activates the brake pulse generator, producing a positive braking pulse at output pin 3. The braking pulse is inverted by Q4D8 and applied to the F/R line of capstan motor drive. This reverses capstan motor rotation, bringing the motor to an immediate halt. Once the motor is halted, the stop detector within IC4A5 terminates the braking pulse.

The high on the SL-OK line is also applied to pin 15 of IC4A5, turning on the slow speed command circuit (which is also triggered by the drum FF pulses). As discussed in Sec. 6-3, different video heads are used for 2H and 6H (2H-L'LR' and 6H-RR'L). As a result, the phase of the drum FF signal in 2H must be 180° out of phase with that in 6H to ensure that the correct video heads are tracking the tape during the individual run-coast-brake-still segments of slow motion. The drum FF phase reverse is done by the XOR gate within IC4A6.

The drum FF signal is applied to one input of the XOR in all modes. In 2H, a high is applied to the other input of the XOR, so the XOR functions as a simple inverter, inverting the drum FF signal. In 6H, pin 2 of IC4A6 is low, and the drum FF signal appears at the output of the XOR in the original, noninverted form. The drum FF signal from the XOR is inverted by Q4F0, and forms into a sawtooth signal by C4E6. The resultant sawtooth is applied to the slow speed command input at pin 10 of IC4A5.

When the rise of the sawtooth reaches a predetermined level, the slow speed command activates the drive pulse generator which, in turn, develops logic for a number of applications, as follows:

1. An output pulse is generated at pin 6, IC4A5, and directed to the gate array IC8A7 in the head-switching circuit. This turns on the appropriate video heads to be used in a given slow-motion mode, as discussed in Sec. 6-3.
2. The pulse at pin 6, IC4A5, is also directed to the drum motor drive through D4D1. This increases the drum speed slightly during the slow-motion run segment.
3. A pulse is directed (within IC4A5) to the capstan drive generator, which responds by producing a pulse at pin 9. This output drives the capstan motor (through D4D0) in a forward direction.
4. An additional signal within IC4A5 disables the coast pulse generator, resulting in a high at pin 8 during the run segment. This high is inverted to a low by Q4C0, holding Q4B9 on and applying 24 V to the capstan drive.
5. The CTL inhibit MM is enabled, inhibiting the slow-tracking MM and negating the effect of any CTL pulses occurring during the active run segment. The duration of the inhibit is determined by the time constant of R4M5 and C4G4 at pin 21.

The duration of the drive pulse at pin 9, and the duration of the run segment of slow motion, is determined by the adjustable time constant of the network at pin 5. In 6H, the high on the 6H line enables Q4D6, which effectively shorts across VR4A9, the 2H slow adjustment. The time constant in 6H is then set by C4F9 and VR4B0, the 6H slow adjustment. In 2H, Q4D6 is disabled by a low on the 6H line, VR4A9 is made part of the circuit, and the time constant is determined by C4F9, VR4B0, and VR4B9. Note that since the 6H slow adjustment is an active part of the circuit in both 2H and 6H, the 6H slow adjustment should always be performed first.

Slow-motion speed is set by controlling the point at which the drum sawtooth (pin 10) triggers, or sets, the slow speed command circuit. Once the speed command is set, the circuit cannot be triggered again until reset. This point is determined by the voltages at pins 16 and 17 of IC4A5. Pin 16 accepts the slow-VR voltage developed by system control, at an amplitude dependent upon the setting of the slow-motion speed controls. When the speed command is reset, pin 17 drops to a low, rapidly discharging C4G2. When the drum sawtooth sets (or triggers) the speed command, the low at pin 17 is removed and C4G2 begins to charge through R4M1.

The time constant of R4M1 and C4G2 is chosen to be relatively long. When the charge on C4G2 reaches the value of the slow-VR voltage at pin 16, the slow speed command is reset and can be triggered by the next recurring sawtooth. Since the time constant at pin 17 is fixed, the slow-VR voltage at pin 16 determines the point at which the slow speed command is reset. In turn, this determines the repetition rate of the slow-motion cycles and the effective slow-motion speed.

As an example, a lower slow-VR voltage at pin 16 results in a faster reset of the speed command and an increase in the slow-motion speed. Conversely, a high slow-VR voltage delays the reset time, and the slow-motion speed is decreased.

9-2.2 Coast segment

When the drive pulse is terminated at the end of the run segment, as established by the time constant at pin 5, the coast pulse generator is enabled, generating a negative-going pulse at pin 8 of IC4A5. Following inversion by Q4C0, the resulting positive-going pulse disables Q4B9, removing the 24 V from the capstan motor drive, allowing the capstan motor to coast without power. (Compare this slow-motion circuit with that described in Sec. 8-4.)

Termination of the drive pulse simultaneously disables the CTL inhibit MM which, in turn, enables the slow-tracking MM. When a CTL pulse is present at the slow-tracking MM input, pin 13, the MM is triggered. This generates a pulse the width of which is determined by the time constant of the network at pin 1 of IC4A5. The time constant (and thus the pulse width, and the duration of the coast segment) are adjusted by the 6H preset tracking control VR4A7, the 2H preset tracking control VR4A8, and the slow tracking control VR8A1.

In 6H, VR4A8 is effectively removed by Q4D3 (which is turned on by the 6H command). In 2H, Q4D3 is turned off, returning VR4A8 to the circuit. Since the 6H preset tracking control VR4A7 is in the circuit for both 2H and 6H, adjustment of the 6H preset tracking should always be performed first.

When the slow-tracking MM is triggered by the incoming CTL pulse, pin 1 attempts to go low, but cannot do so until C4F6 charges through the tracking controls. When the voltage at pin 1 drops to 4.5 V, the slow-tracking MM is reset and the output is terminated. This turns on the brake pulse generator, stopping the coast segment, and initiates the brake segment.

9-2.3 Brake segment

When the brake pulse generator is turned on at the end of the coast segment, a signal is generated. This signal disables the coast pulse generator, producing a high at pin 8 of IC4A5. This high turns on Q4B9, through Q4C0, and supplies 24 V to the capstan motor drive. The signal also turns on the capstan drive generator within IC4A5, producing a capstan drive signal at pin 9 of IC4A5. Simultaneously, the brake pulse generator produces a high at pin 3.

This high is inverted to a low by Q4D8, driving the F/R line low and reversing direction of the capstan motor (to produce braking).

FG pulses from the capstan motor are applied to the stop detector input at pin 19 and, through the stop detector, are also applied to C4F7 at pin 2. So long as the capstan motor is rotating, the frequency of the FG signal is sufficiently high to maintain C4F7 in a discharged condition. The negative-going portions of the FG signal rapidly discharge C4F7, and the positive-going portions of the FG signal are too short to allow C4F7 to develop a charge. As the capstan motor approaches a stopped condition, the frequency of the FG signal is reduced sufficiently to allow C4F7 to charge during the positive portions of the signal.

When the charge across C4F7 reaches 4.5 V at pin 2, the stop detector produces an internal pulse which resets the brake pulse generator. This terminates the braking period and initiates the still segment until the slow speed command is once again triggered by the incoming drum sawtooth, and the entire slow-motion cycle is repeated.

If, for some reason, the stop detector malfunctions and does not sense the stop point of the capstan motor, the brake segment would not be terminated, and the capstan motor could rotate in reverse. A backup brake reset circuit is used to prevent this condition.

When the brake pulse generator is activated, at the end of the coast segment, pin 4 of IC4A5 attempts to go low, but cannot do so until C4F8 charges through R4L1 (in 6H) or R4L2 (in 2H). When the voltage at pin 4 has decreased to 4.5 V, the brake pulse generator is reset and the brake segment of slow motion is terminated, preventing reverse rotation of the capstan motor.

9–2.4 Still segment

Figure 9–5 shows the basic still circuit. The still function is essentially an extension of slow motion, with most of the still circuits located within slow-motion IC4A5. When still is first selected, the slow-motion circuit initiates three complete run-coast-brake-still cycles prior to still. If the slow-motion circuit is properly adjusted, with noise-free slow-motion pictures, then a noise-free picture can be expected in the still mode (automatically).

When still is selected, a high is placed on the ST-OK (still-OK) line and is applied to Q4E7. The emitter goes high immediately, but the base remains low momentarily due to the charging action of C4F3. Since Q4E7 is PNP, and the base voltage is lower than the emitter voltage, Q4E7 conducts until C4F3 develops a charge sufficient to disable Q4E7. The value of C4F3 is chosen to allow conduction of Q4E7 for a period of one second. This drives the still input, pin 14 of IC4A7, high.

The high at pin 14 turns on Q4E3, clamping the slow-VR (slow-speed) voltage at pin 16 to 1.4 V, representing the fastest slow-motion speed. The high at pin 14 also turns on the slow-speed command and the slow/still discriminator. In turn, the slow/still discriminator enables the still timer.

Slow Motion (Continuously Variable)   215

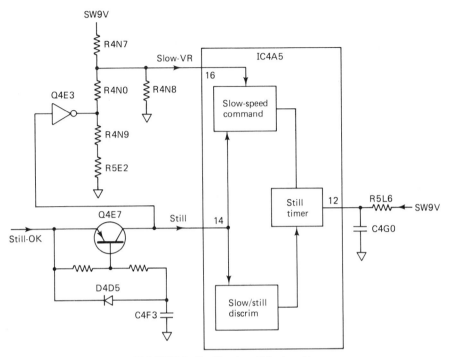

FIGURE 9-5   Basic still circuit

The timing of the still timer is set by the time constant of the component network at pin 12. The values of R4L6 and C4G0 are chosen so that the still timer outputs an internal disable signal, following completion of three complete slow-motion cycles. This disables the slow-speed command and maintains the still segment of the *last* slow-motion cycle until such time that the still mode is released and normal servo control is restored.

In the event that CTL pulses are missing from the tape, the brake pulse generator cannot be enabled during the three slow-motion cycles. In that event, the brake segment cannot be initiated, the circuit remains in coast, and the capstan motor would merely coast to a stop. The time constant of the components in the base circuit of Q4E7 prevent such an occurrence. (Conduction of Q4E7 occurs after one second.) This removes the high at pin 14 of IC4A5 and stops the capstan motor by turning on the internal brake pulse generator.

9-2.5   Slow motion troubleshooting/repair

Troubleshooting the slow-motion circuits shown in Figs. 9-4 and 9-5 is mostly a matter of checking inputs and outputs using an oscilloscope or logic probe (as described for speed search in Sec. 9-1.6). If the inputs to IC4A5 are absent or abnormal, trace the lines back to the source (system control in most cases).

216 Special Effects

If the inputs are good, but the outputs are bad, suspect IC4A5. Of course, before you plunge into slow-motion circuits, make sure that the problem is in slow motion and not something common to other modes (such as a failure in the capstan servo, described in Chapter 8, or the drum/cylinder servo, described in Chapter 7).

9-3 MECHANICAL BRAKE

Figure 9-6 shows the mechanical brake circuits. Mechanical braking is used in addition to the electronic braking (described in Sec. 9-2) to increase the precision of braking action. The same mechanical brake is used in both 2H and 6H. In either still or slow motion, Q4C1 is enabled by a high at the base from either the ST-OK or SL-OK lines, respectively, via OR-gate D4A6. Turning on Q4C1 activates the brake relay K4A1, switching the contact from the 24-V terminal to ground terminal.

Relay K4A1 grounds pin 2 of the CN plug and energizes brake solenoid T471. This applies braking to the capstan motor. The abrupt change on pin 2 of plug CN (from 24 V to ground) is coupled to Q4C2 through C4E9. Q4C2 conducts momentarily, turning on Q4C3 and supplying 24 V (momentarily) to the opposite end of the brake solenoid. This provides a much higher, yet momentary, braking force during the mechanical braking period.

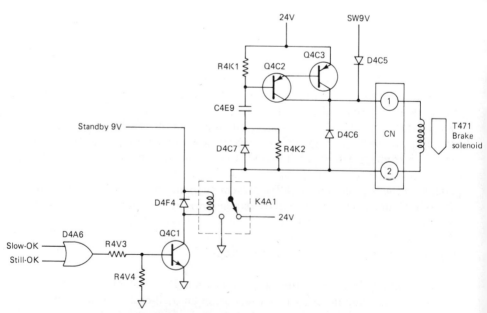

FIGURE 9-6 Mechanical brake circuits

When C4E9 charges to the threshold of Q4C2, both Q4C2 and Q4C3 are disabled, removing the 24 V from pin 1 of plug CN. At that time, 9 V is applied to the brake solenoid, through D4C5, to serve as a holding voltage. When still or slow motion is terminated, the high at the base of Q4C1 drops to a low, disabling Q4C1 and deenergizing relay K4A1. The K4A1 contact switches to the 24-V terminal, and the brake solenoid is deenergized, releasing the mechanical brake.

9-3.1 Mechanical brake troubleshooting/repair

Again, troubleshooting the mechanical brake circuits is a matter of checking inputs and outputs. Of course, it is possible that the circuits may be good but the brake is not properly adjusted. So, as a start, check the service literature for any mechanical brake adjustment procedures. (Do this right after you have checked to make sure that the VCR you are servicing has mechanical braking. Many VCRs do not.)

If you cannot cure the problem by adjustment, check that the brake solenoid is energized when slow and/or still modes are selected. If not, trace back to the source. For example, first check for a high on either the SL-OK or ST-OK lines when slow and/or still are selected. If missing, suspect system control. If the slow and/or still commands are good, check that relay K4A1 is turned on. (Pin 2 of plug CN should be at ground if K4A1 is actuated.) If K4A1 is not turned on, suspect D4A6 and Q4C1.

If K4A1 is actuated, check for about 9 V at pin 1 of plug CN. If the 9 V is missing, suspect D4C5. Next, monitor pin 1 of plug CN while switching from normal play to still or slow. Pin 1 of plug CN should go from about 9 V to near 24 V, momentarily, when either slow or still is selected. If not, suspect Q4C2, C4C3, C4E9, D4C7, R4K2, D4C6, and R4K1.

If the voltages are correct at pins 1 and 2 of plug CN, but the brake is not actuated when still or slow are selected, suspect the brake solenoid T471.

9-4 QUASI-SYNC GENERATOR (SUBSTITUTE VERTICAL SYNC)

In many modern VCRs that have special effects (most do), the noise bars generated in the various special-effect modes are forced into the vertical blanking interval of the TV display. This is to produce noise-free pictures during slow, still, etc. Unfortunately, a very undesirable side effect can be produced by this configuration. Since the vertical sync pulse of the TV display also occurs within the vertical blanking interval, the introduction of noise can affect the vertical sync pulse. This often produces *vertical jitter* and can result in *vertical instability*.

In some modern VCRs, the problem of vertical instability is eliminated

218 Special Effects

by developing a *substitute vertical sync pulse* for use exclusively during special-effect modes, thus eliminating dependency on the recorded vertical sync pulses. Figure 9-7 shows the circuit for generating such a substitute pulse. (This circuit is called the quasi-sync generator in VCR of Fig. 1-1a.) Figure 9-8 shows the related circuit waveforms.

The quasi-sync signal is developed using both excursions of the 30-Hz drum/cylinder FF. As discussed in Chapter 7, the drum/cylinder FF signal is phase-locked during playback to a locally generated 60-Hz signal taken from a crystal-controlled oscillator, and is a faithful reproduction of the vertical sync frequency.

The 30-Hz drum FF signal is applied to the junction of R4B6 and R4B7, through Q4A0, and then to IC4A0. The rise and fall of the pulses are delayed slightly before application to IC4A0. The delay introduced to the fall of the FF pulse is set by R4B6, VR4A7 (the prestill adjustment), D4A0, and C4A0. The delay introduced to the rise of the FF pulse is set by R4B7, R4A3, D4A1, and C4A0. The delayed FF pulses are applied to an XOR gate within IC4A0.

When the inputs to the XOR gate are identical, the XOR output is low. If either input varies from the other, the XOR output goes high. Because of the delay introduced by R4A4 and C4A1, making the pin 2 input different from the pin 1 input, the XOR output is high during both the rise and fall of the input FF pulse. In effect, the XOR functions as a type of frequence doubler, accepting a 30-Hz drum FF signal and generating a 60-Hz output, as shown by the waveforms in Fig. 9-8.

The output of the XOR provides one input to the NAND gate within IC4A0. The other NAND gate input receives a $\overline{N}$ voltage (generated within IC4A0). The $\overline{N}$ voltage is high during all special-effects modes, so the NAND

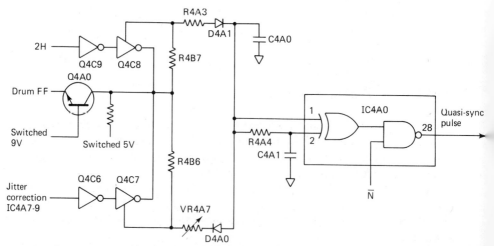

FIGURE 9-7 Quasi-sync generator circuits

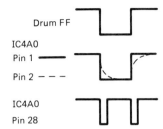

FIGURE 9-8 Quasi-sync generator waveforms

gate is enabled only during special effects. (When $\overline{N}$ is low during other modes, the output of the NAND gate remains high.)

The NAND gate functions as an inverter, producing two negative-going quasi-sync output pulses for each drum FF pulse. (One output pulse is developed during the rise, and the other pulse is developed during the fall, of each input drum FF pulse.) The quasi-sync output is thus generated at a 60-Hz rate.

The quasi-sync generator of Fig. 9-7 also includes components which further minimize vertical jitter (during special effects) with changes in tape speed. For example, in 2H, timing of the quasi-sync pulse generated by the rise of the drum FF pulse is accelerated slightly. When 2H is selected, Q4C9 and Q4C8 are turned on, shorting across R4B7, and decreasing the time constant of the delay circuit. This reduces the delay time of the rise signal applied to pin 1 of IC4A0, so the quasi-sync pulse occurs slightly earlier in 2H than in 6H.

Additional jitter correction is required in 6H, due to the very low tape speed. During the run segment of slow motion (Sec. 9-2.1), a jitter correction pulse from pin 9 of IC4A7 turns on Q4C6 and Q4C7, shorting across R4B6, and decreasing the delay of the quasi-sync pulse developed from the fall of the drum FF pulse. As a result, the sync pulses (during the run segment of 6H) are accelerated slightly to reduce vertical jitter at the extremely slow tape speed.

9-4.1 Quasi-sync generator troubleshooting/repair

Troubleshooting the quasi-sync generator circuits is, as usual, a matter of checking inputs and outputs. As a start, check the service literature for any adjustment procedures, such as adjustment of VR4A7, once you suspect a problem in the quasi-sync. (Keep in mind that the quasi-sync circuits are used to *minimize vertical jitter during special effects*. If you have vertical jitter in all modes, the problem is probably not in the quasi-sync circuits.)

If you have vertical jitter in special effects, and you cannot cure the problem with adjustment, check for proper drum FF pulses at the input of Q4A0, and for proper quasi-sync pulses at the output of IC4A0 (Fig. 9-8). If you have good drum FF pulses, but absent or abnormal quasi-sync pulses, trace through the Fig. 9-7 circuit with an oscilloscope or logic probe. For example, if the drum FF pulses are absent or abnormal at the junction of R4B6/R4B7, suspect Q4A0.

If you get vertical jitter in 2H, but not 6H, suspect Q4C8/Q4C9. Also check for a 2H command from system control at the input of Q4C9. If you get vertical jitter in 6H only, suspect Q4C6/Q4C7. Also check for a jitter-correction pulse at the input of Q4C6.

10

TUNER AND FREQUENCY SYNTHESIS CIRCUITS

This chapter is devoted to troubleshooting and repair of VHF/UHF tuners found in modern VCRs. The tuners used in early-model VCRs are similar to those of early-model TV sets (and include the usual solid-state mixer/local-oscillator combination). The tuners of most modern VCRs use a microcomputer-controlled frequency synthesizer (FS). This provides convenient push-button channel selection with automatic channel search and automatic fine-tune (AFT) capability.

The key element in any FS system is the phase-lock-loop (PLL), which controls the variable-frequency local oscillator of the tuner, as required for channel selection. So we start our discussion with PLL basics. (Note that the PLLs used in VCRs are essentially the same as those used in the FS tuners of modern TV sets.)

10-1 PLL BASICS

Figure 10-1 shows the basic PLL circuit. PLL is a term used to designate a frequency-comparison circuit in which the output of a variable-frequency oscillator (VFO) is compared in frequency and phase to the output of a very stable (usually crystal-controlled) fixed-frequency reference oscillator. Should a deviation occur between the two compared frequencies (or should there be any phase difference between the two oscillator signals), the PLL detects the degree of frequency error and automatically compensates by tuning the VFO up or down in frequency until both oscillators are locked to the same fre-

222 Tuner and Frequency Synthesis Circuits

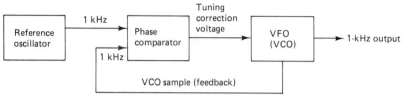

FIGURE 10-1 Basic PLL circuit

quency and phase. (The loop is said to be *locked* at this time.) The accuracy and frequency stability of a given PLL circuit depends upon the accuracy and frequency stability of the reference oscillator (and on the crystal that controls the reference oscillator).

In the basic PLL circuit of Fig. 10-1, the VFO is a form of voltage-controlled oscillator (VCO) with a desired output frequency of 1 kHz. The actual output frequency of the VCO depends on the tuning voltage produced by the phase comparator and applied to the VCO. As shown, the phase comparator receives two input signals, both at 1 kHz. One signal is from the reference oscillator, while the other signal is a sample taken from the VCO.

Any frequency or phase variation in the output of the VCO, when compared to the stable 1-kHz reference frequency, causes the phase comparator to produce a correction voltage. The magnitude of the correction voltage depends on the amount of frequency/phase deviation. The polarity of the correction voltage depends on the direction of frequency/phase variation. The correction voltage is applied to the VCO as an increase or decrease in the tuning voltage. Changes in the tuning voltage change the VCO frequency as necessary to make the VCO output of the same frequency and phase as the reference oscillator output. When this occurs, the tuning voltage stabilizes and the PLL is said to be *locked in.*

Figure 10-2 shows a somewhat more sophisticated PLL circuit, one capable of comparing frequencies that are *not identical* in frequency. The circuit of Fig. 10-2 includes a *divide-by-10 element,* which divides the VCO frequency by 10 prior to application to the phase comparator, and a *low-pass filter,*

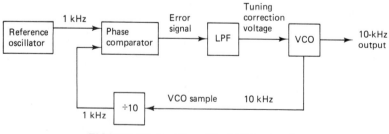

FIGURE 10-2 Simplified PLL circuit

which acts as a buffer between the phase comparator and the VCO. Note that while the inputs to the phase comparator remain at 1 kHz when the loop is locked, the output frequency of the VCO is now 10 kHz due to the action of the divide-by-10 element.

Figure 10-3 shows a PLL circuit more similar to that found in modern VCRs and TV sets, but far less complex. The system is generally called an "extended" PLL, and holds the local-oscillator frequency to some harmonic (or subharmonic) of the reference oscillator. When the loop is locked, the PLL maintains a *fixed phase relationship* between the local oscillator and the reference oscillator. Under locked conditions, the local-oscillator frequency is held to that of the crystal-controlled reference.

The fixed divide element of Fig. 10-2 is replaced by a variable divider ($\div N$) in Fig. 10-3. This variable divider is programmed by 4-bit data from the microcomputer to divide the VCO output frequency by a specific number. The variable divider makes possible many local-oscillator frequencies. A fixed divider is used between the reference oscillator and phase comparator to reduce the reference-oscillator frequency to a more workable, easily compared frequency, while still maintaining maximum frequency range. A channel change is accomplished by varying the division ratio of the programmable divider ($\div N$) with 4-bit data commands from the microprocessor. (Note that the terms *microcomputer* and *microprocessor* are used interchangeably throughout VCR literature, so we do the same, even though there is a technical difference.) For accurate tuning of a given channel, the division steps of the programmable divider are made small, and a low-frequency reference to the comparator (976.6 Hz) is used.

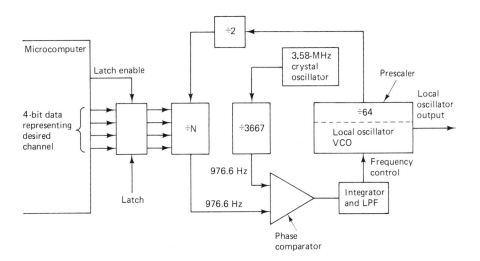

10-1.1 Pulse swallow control

Many (but not all) PLL tuning systems use some form of pulse swallow control (PSC), such as shown in Fig. 10-4. PSC allows the division ratio of the programmable divider to be changed in much smaller steps. This produces a much faster reaction to a channel change command. The large steps in division ratio also allow the use of a higher reference frequency at the comparator, simplifying design requirements of the low-pass filter.

The PSC system uses a very high-speed prescaler with a variable division ratio (which contrasts to the fixed division ratio prescaler of Fig. 10-3). The variable division ratio of the prescaler can be altered as required to produce the more subtle changes in frequency needed for optimum channel tuning.

The PSC signal is a series of positive pulses. As the number of pulses on the PSC line increases, the division ratio of the prescaler also increases. The number of pulses on the PSC line, when a given channel is selected, is determined by the microprocessor for each channel position. As a result, the overall division ratio for a specific channel is the prescaler division ratio multiplied by the programmable divider division ratio. The result of division at any channel is a 5-kHz output frequency to the comparator when the tuner local oscillator is tuned to a given channel frequency.

The programmable divider determines the MSD (most significant digit) of the overall division ratio, and the prescaler determines the LSD (least significant digit). As a result, the programmable divider determines the basic channel frequency, and the prescaler performs the fine adjustments to the channel frequency required for optimum tuning.

If the PSC signal is missing, the division ratio of the prescaler remains fixed at some static level, and only large changes in division of the local oscillator (via the programmable divider) are possible. In automatic-tune mode, with a missing PSC output, the AFT circuit attempts to alter the tuning by

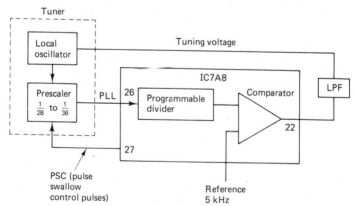

FIGURE 10-4 Pulse-swallow control (PSC) circuits

varying the division ratio of the divider. Since the division ratio can only be changed in large steps, an optimum tuning point is never reached. So the AFT continually changes the division ratio, searching for the optimum tuning point. The system remains on the selected channel position, but continues to scan through the channel position due to AFT action. Under these same conditions, in manual-tuning mode, the manual fine-tuning adjustments can only make large changes in the division ratio and optimum tuning cannot be attained. *So if you are troubleshooting any tuner with PSC, and you cannot tune in a channel with AFT or manual, check the PSC line (pin 27 of IC7A8, in this case) for pulses.*

10-2 TYPICAL FS TUNING SYSTEM

Figure 10-5 is the block diagram of the FS tuning system for a typical VCR (that shown in Fig. 1-1a). This FS tuning system has the following capabilities:

1. 139-channel (including cable) tuning capabilities (VHF, UHF, mid, super, and hyper bands, including MID MAND subchannels A-1/A-5 and 5A).
2. AFT pull-in range of ±2.4 MHz.
3. Positive channel scan, scanning 2.4 MHz above and below the normal carrier frequency of each channel, attempting to locate a signal.
4. Tuner B+/band-switching and low-pass filter are combined into a single IC.

The system is controlled by a microprocessor IC7A0, which we designate as the UCP in this discussion. Power for the UCP is applied to pins 14 and 16. Timing for the UCP is provided by a 4.5-MHz crystal-controlled master clock oscillator at pins 17 and 18.

The display digit drive logic originates from the PD2 and PD3 outputs, pins 25 and 26, with segment data output from the Sa-Sg terminals, pins 1-7.

Pins 1-7 also provide *scanner* signal outputs to the front-panel *keyboard matrix*. These signals continually scan the keyboard until a switch closure is detected. At that point, the appropriate scanner signal is applied to one of four I/O ports PB0-PB3, pins 9-13, causing the UCP to respond by producing the necessary outputs.

The FS system also features *channel-scan* capabilities, bypassing channels with no signals and stopping automatically on active channels. The channel-scan feature includes a *sync-detection* system, which detects the presence of horizontal sync when an active channel is scanned. When horizontal sync is detected, sync-detect IC121 produces a high at pin 6. This high turns

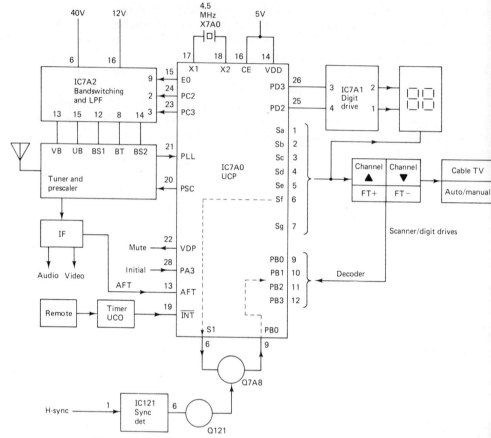

FIGURE 10-5 Typical FS tuning-system block diagram

on Q121-Q7A8 and applies an Sf scanner signal to the PB0 decoder input. The UCP responds by terminating channel scan and initiating AFT.

The sample AFT voltage, taken from the IF (intermediate frequency) circuits, is applied at pin 13. When the AFT voltage is within the range of 1.72 to 3.28 V, the UCP terminates AFT action.

Channel change commands from the remote control preamp, via the timer microprocessor (Chapter 14), are applied to the UCP at pin 19, the INT input. Pin 19 also accepts channel data from the timer microprocessor when in the reserve program mode.

Pin 28 of the UCP is used in the initializing (resetting) process when power is first applied to the VCR. The initializing process is discussed in Sec. 10-2.9.

The output at pin 22 goes low during a channel change. The low is used to mute the audio (Chapter 11) so that no unwanted noise occurs during a channel change.

The output at pin 20, the pulse swallow control or PSC, applies a series of pulses to the prescaler section of the tuner, as described in Sec. 10-1.1. The pulses select the division ratio of the prescaler. The tuner local-oscillator signal is divided by the prescaler and then applied to the PLL, pin 21, where the signal is again divided by the UCP programmable divider. The number of pulses at the PSC output, and the division ratio of the UCP programmable divider, *have specific values for each channel,* predetermined by UCP programming for the selected channel.

If the tuner local-oscillator frequency drifts, an error correction output (EO) voltage is provided at pin 15. The amplitude of EO depends on the amount of frequency error. The EO voltage is applied to band switching and LPF IC7A2, which produces a corresponding tuning voltage (BT) applied to the tuner. The BT voltage returns the local oscillator to the proper frequency for the channel selected.

IC7A2 also serves as the band-switching/tuner B+ source, accepting input logic at pins 2 and 3 from pins 23 and 24 of the UCP. This develops the appropriate output voltages required to power the tuner, as discussed in Sec. 10-2.3.

10-2.1 Tuner/prescaler

The tuner shown in Fig. 10-5 is a UHF/VHF combination which receives operating voltages at eight input terminals simultaneously. Two input voltages are *fixed,* and should be present at all times. These fixed inputs include BP, a B+ supply input for the prescaler, derived from the switched 5-V supply; and VM, a B+ supply for the mixer, derived from the 12-V supply.

Two of the input voltages are *variable*. These variable inputs include U-AGC, the RF AGC input for both UHF and VHF modes, the voltage value depending on the signal strength of the channel being received; and BT, the tuning voltage input, a variable dc voltage at a *specific value for each channel,* derived from the low-pass filter in IC7A2.

The four remaining input voltages to the tuner are *switched* and depend on the mode (VHF, UHF, CATV) and the band within the mode (low/high, VHF, mid, super, or hyper band) that is selected. The switch voltage include VB, the VHF tuner B+ input which powers the VHF tuner in both VHF and CATV modes, derived from pin 13 of IC7A2; UB, the UHF tuner B+ input which powers the UHF tuner in the UHF mode, derived from pin 15 of IC7A2; BS1 and BS2, band-switching voltage inputs which select the range of frequencies that can be tuned within a given mode, derived from pins 12 and 14 of IC7A2.

10-2.2 Key matrix

Figure 10-6 shows the basic key matrix circuit. Scanner signals Sa-Sg at pins 1-7 of IC7A0 are applied to the key matrix pad assembly. The output of the key matrix is applied to the four decoder inputs of IC7A0 at pins 9-12. When a button on the key matrix is pressed, a specific scanner signal is applied to a specific decoder input, and the UCP reacts by performing the desired operation.

For example, if the *channel-up* button is pressed, with the MAN/AUTO switch in MAN position, an Sd scanner signal (pin 4) is applied to the PB0 input (pin 9), and the UCP automatically performs the necessary operations to change the system to the next higher channel.

Note that the CABLE/TV and MAN/AUTO switches are integral parts of the matrix and alter operation of the UCP to correspond with the tuning mode selected. In TV, an Sg signal is continuously applied to the PB0 input. In CABLE, the Sg input is removed from the PB0 input. In MAN, the Sg signal is continuously applied to the PB3 input. In AUTO, the Sg signal is removed from the PB3 input.

Q7A9 is turned on for 20 ms during the initialization process, as described in Sec. 10-2.9. This applies an Se signal to the PB1 input, enabling the $\overline{\text{INT}}$ input (pin 19 of IC7A0, Fig. 10-5) to accept signals originating from the timer processor (Chapter 14) or remote control (Chapter 4).

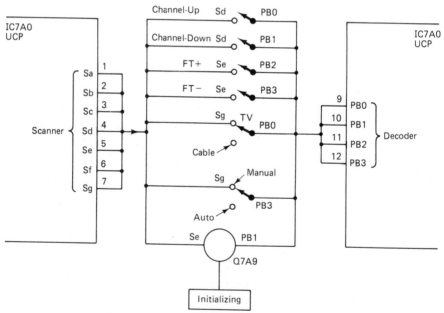

FIGURE 10-6 Basic key matrix circuit

10-2.3 Tuner B+ and band switching

Figure 10-7 shows the basic band-switching circuits. Band-switching information from IC7A0 is automatically output (at pins 23 and 24) for each channel selected, and applied to IC7A2, which develops the band-switching and tuner B+ voltages applied to the tuner. The band-switching outputs from IC7A2 are *three-state,* with one state being *high-impedance,* similar to the outputs used in most digital ICs. Figure 10-8 shows the equivalent circuit for a high-impedance output. When switch S is closed, the output terminal is a conventional two-state device, either high or low. When S is open, the output terminal is isolated from the internal circuits of the IC by R, a very large impedance. When in the high-impedance state, the output terminal can be considered essentially disconnected from the external circuits.

The truth table for the band-switching circuit is shown in Fig. 10-7. The Z notation at the outputs of IC7A2 denotes the high-impedance or high-Z state. For example, in the UHF band, IC7A0 produces lows at both pins 23 and 24. The two lows are applied to pins 2 and 3 of IC7A2, which responds by generating a high at pins 12 and 15, while simultaneously switching pins 13 and 14 to the high-impedance state.

The high at pin 15 is applied to the UB tuner input, supplying UHF B+ to the tuner. The high-Z state of pin 13 removes VHF B+ from the VB tuner

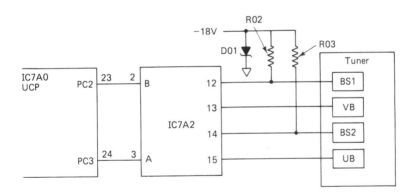

Band	IC7A0 pins		IC7A2 pins				Tuner voltages			
	23	24	15	14	13	12	UB	BS2	VB	BS1
UHF	L	L	H	Z	Z	H	+12	−12	0	+12
SUP/HYP	H	L	Z	H	Z	H	0	+12	+12	+12
VHF-L	L	H	Z	Z	H	Z	0	−12	+12	−12
MID/VHF-H	H	H	Z	Z	H	H	0	−12	+12	+12

FIGURE 10-7 Basic band-switching circuits

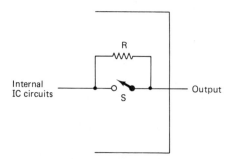

FIGURE 10-8 Equivalent circuit for high-impedance output

inputs, and the high-Z at pin 14 removes positive voltage from the tuner BS2 input. This allows BS2 to go negative (via R03 and the -18-V supply). The high on BS1, and the negative on BS2, switches the tuner to the UHF band.

The use of such truth tables for troubleshooting is obvious. For example, if you cannot receive any UHF channel, check for corresponding inputs/outputs at IC7A0/IC7A2 and appropriate voltages at the tuner. If the outputs at pins 23/24 of IC7A0 are not correct, with UHF selected, suspect IC7A0. If the IC7A0 outputs are good, but the IC7A2 outputs are not, suspect IC7A2. If the IC7A2 outputs are good, but you cannot get any UHF channel, suspect the tuner. (Now if you can devise some method to make the manufacturers put such truth tables in the service literature, life will be much easier!)

10-2.4 Channel scan/AFT

Figure 10-9 shows the channel-scan sequence. When the MAN/AUTO switch is set to AUTO, and the channel-up or channel-down button is pressed, the UCP begins a sequential scanning process, incrementing from channel to channel, up or down, bypassing inactive channels, and stopping automatically when a signal is detected. As each channel position is reached, a secondary scanning process follows. The UCP automatically scans 2.4 MHz above and below the nominal carrier frequency for the channel, attempting to find a signal. If no signal is found, the UCP moves to the next channel position in sequence. As a result of the secondary scanning process, channel scan in the tuning system described here is somewhat slower than that of other FS systems. However, the effectiveness is measurably increased, especially when *offset carriers* (such as used in many CATV systems) are involved.

If no signal is detected at any channel position, the system continues the scanning process indefinitely. Once the system has come to rest on an active channel, channel-scan is terminated and AFT is used to ensure optimum tuning. Pressing either channel-scan button with the MAN/AUTO switch set to MAN causes the system to increment, up or down, one channel each time a channel-scan button is pressed.

The AFT circuit is not active in the MAN mode. Fine tuning, if required,

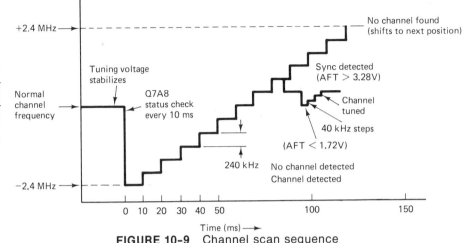

FIGURE 10-9 Channel scan sequence

must be done by the FT+ and FT− buttons on the front panel. Pressing either button causes the UCP to increment frequency in steps of 40 kHz to the maximum range of 2.4 MHz above or below the normal frequency of a given channel. Fine-tuning variations are not entered into the IC memory, and adjustments required on a given channel must be repeated each time the channel is selected.

Note that fine tuning, even in the MAN mode, is generally not required unless the broadcast source is off-frequency, as is sometimes the case with certain CATV systems, video games, and so on. The FT+ and FT− buttons are not active in the AUTO mode.

10-2.5 Channel scan

Channel-scan operation is controlled, in part, by the sync detector shown in Fig. 10-10. During channel scan, the UCP automatically increments from one channel to the next by altering the value of N (division factor by which the tuner local-oscillator frequency is divided), as discussed in Sec. 10-1. When the division factor is changed, the tuning voltage applied to the tuner is also changed to establish a "locked" condition of the PLL at the appropriate local-oscillator frequency required for the new channel.

There is a slight delay until the tuning voltage has stabilized at the value required for the new channel. The UCP then determines the status of Q7A8 shown in Fig. 10-10. If Q7A8 is off (no Sf signal present), AFT is inhibited, and the UCP automatically alters the tuning voltage to shift the local-oscillator frequency downward by 2.4 MHz. This occurs during both the channel-up and channel-down scanning processes.

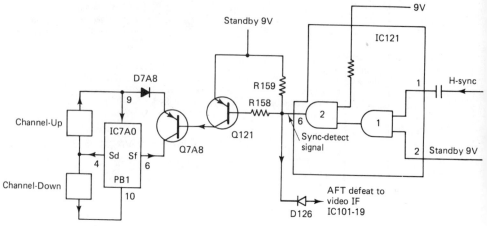

FIGURE 10-10 Sync detector circuit for channel scan

The status of Q7A8 is checked once again. If Q7A8 is still off, the UCP shifts the local-oscillator frequency upward in 240-kHz steps until the local-oscillator frequency is 2.4 MHz above the nominal carrier frequency. As shown in Fig. 10-9, steps occur each 10 ms, and the status of Q7A8 is checked at each step. If Q7A8 remains off for the duration, the system determines that no signal exists at any point within the tuning range of that particular channel and moves to the next channel position, up or down, where the process is repeated.

If at any time during the process an active signal is found, as shown by Q7A8 turning on and passing the Sf signal from pin 6 to pin 9 of IC7A0, the UCP responds to AFT signals at pin 13 (as shown in Fig. 10-5). AFT action is inhibited as long as Q7A8 is in an off state. Due to the nature of the scanning process, and the fact that the search within a given channel position begins at 2.4 MHz below the nominal broadcast frequency for the channel, the entire range of each channel position is scanned (± 2.4 MHz).

Referring to Fig. 10-9, it can be seen that a channel is considered to be properly tuned if the AFT voltage at pin 13 of the UCP is between 1.72 V and 3.28 V. During the scanning process, once it has been determined that a signal exists (Q7A8 on), the UCP continues to increment, in 240-kHz steps, until the AFT voltage at pin 13 is less than 1.72 V, passing through the optimum tuning point. The IF frequency is too high at this point, and the UCP automatically begins to return to the optimum tuning point in 40-kHz steps, stopping within the range of 1.72 to 3.28 V.

Note that not all FS tuning systems follow this exact sequence, nor do the scanning steps cover the same frequency range. However, a similar process is used for scanning by most FS systems, as discussed in Sec. 10-3.

10-2.6 Tuner sync detection

As indicated in Fig. 10-10, the tuner UCP is instructed to initiate channel scan when an Sd signal from pin 4 is applied to the PB0 input at pin 9 (channel-up), or the PB1 input at pin 10 (channel-down) through closure of the channel-up or channel-down switches, respectively. The command to terminate channel scan is accomplished by applying an Sf signal from pin 6 to pin 9 through Q7A8, when Q7A8 is turned on by inverter Q121 in response to a sync-detect signal from IC121.

IC121 determines the presence of a signal on an active channel by detecting the *horizontal sync pulse* of the signal, and then terminates channel scan by turning on Q7A8. IC121 consists primarily of two AND gates. One input to AND-1, at pin 2 of IC121, is always high because of the 9-V standby voltage. Horizontal-sync pulses are applied to the second input of AND-1, at pin 1. The resulting high from AND-1 turns on AND-2, since the other input to AND-2 is held high by the 9-V supply. The high output from AND-2, at pin 6, turns off Q121 and turns on Q7A8. This applies the Sf scanning signal to the PB0 input, pin 9, of the UCP, and terminates the scanning process.

So long as pin 6 of IC121 is low (inactive channel), D126 is forward-biased, maintaining a low at pin 19 of IC101 (which is the video IF IC). This eliminates AFT during the scan of inactive channels. When pin 6 of IC121 goes high on an active channel, D126 is disabled, allowing normal AFT action.

10-2.7 Tuner LED channel display

Figure 10-11 shows the tuner channel-indicator drive circuits. The tuning system uses a *dynamic* LED channel display. The term *dynamic* indicates a display in which the *individual digits* are activated, one after the other, at a relatively high rate of speed, creating the *illusion of constant illumination* with no detectable flicker.

Digit-enable signals from IC7A0, pins 25 and 26, are applied to digit-drive IC7A1, pins 4 and 3, respectively. The MSD-enable signal from pin 7 of IC7A1 is applied to pin 13 of the LED display V501, and the LSD-enable signal from pin 2 of IC7A1 is applied to pin 14 of V501. Segment-drive signals (Sa-Sg) from IC7A0 are applied directly to V501. Segment drive for an *individual digit* of the two-digit display is applied in parallel, on eight lines. The drive is then switched to the remaining digit of the display in serial form. When the VCR is switched off, there is no longer a switched 9-V supply. This turns on Q7A4 and applies 5 V to pins 3 and 4 of IC7A1, holding both lines high and disabling the digit-enable drive. This turns off the LED channel display.

If the program door of the VCR is opened with the VCR off, the channel display must be turned on to facilitate programming (Chapter 14). When the door is opened, the pro-door switch S8A9 closes, applying standby 9 V to

234 Tuner and Frequency Synthesis Circuits

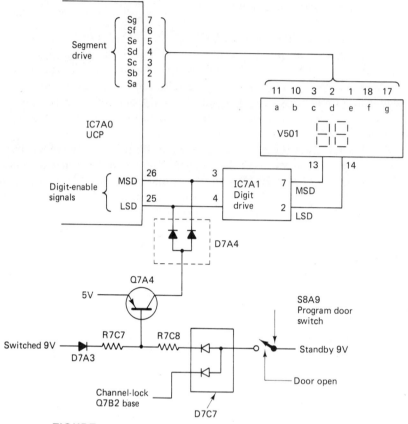

FIGURE 10-11 Tuner channel-indicator drive circuits

Q7A4 (via D7C7), turning off Q7A4. This removes the constant highs at pins 3 and 4 of IC7A1, thus enabling the LED channel display.

10-2.8 Tuner channel-lock

Figure 10-12 shows the tuner channel-lock circuits. To prevent inadvertent channel change in the record mode, or to avoid tampering with programmed channel information, the automatic channel-lock circuit of Fig. 10-12 is used to inhibit a channel-up or channel-down command. The up or down channel-change function is achieved by directing the Sd scanner signal from pin 4 of IC7A0 to the appropriate decoder input of IC7A0 via the key matrix. Transistor Q7B2 is connected in series with the Sd scanner signal, between pin 4 of IC7A0 and the key matrix, and must be conducting for a channel change to occur. Channel lock is accomplished by disabling Q7B2.

Transistor Q7B2 receives switching logic from two sources, both of which

Typical FS Tuning System 235

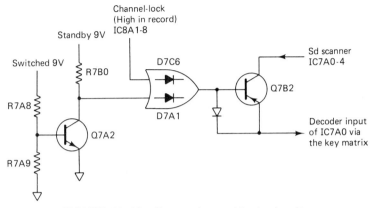

FIGURE 10-12 Tuner channel-lock circuit

are applied through OR-gate D7C6/D7A1. *When the VCR is off,* switched 9-V power is removed, Q7A2 is turned off, and 5 V is applied to Q7B2 via D7A1. This turns off Q7B2 and prevents Sd scanner signals from passing, so there is no channel change. When the VCR is on, the switched 9-V supply turns on Q7A2, which, in turn, turns on Q7B2 to permit channel changes.

During record, the high output from pin 8 of IC8A1 (the timer processor IC, Chapter 14) is applied to Q7B2 through D7C6, turning off Q7B2 and preventing channel change in the record mode.

10-2.9 Tuner initializing (reset)

Figure 10-13 shows the tuner initializing circuits. Figure 10-14 shows the related timing sequence. When power is first applied to the VCR, the tuner microprocessor IC7A0 must be initialized (reset), driving all outputs to zero. If not, erratic operation of the tuner can result. Also, the initializing process must occur in a specific sequence or the remote input at pin 19 will not accept remote-control commands.

When power is first applied, the 5-V line goes high, making pin 14 of IC7A0 high. Simultaneously, pin 12 goes high through circuits within IC7A0. The high at pin 12 is applied directly to the collector of Q7A7, and to the base of Q7A7 through R7D4 and R7D5. The high at the base is delayed for 250 ms by the charging action of C7B2. After 250 ms, the charge across C7B2 is sufficient to turn on Q7A7, which conducts and applies a high to pin 28 of IC7A0 through D7A7. (The first step in initialization requires that pin 14 go high 250 ms before pin 28.)

The high at pin 28 is applied to Q7A9 through R7H2 and R7G1. However, Q7A9 does not conduct for a period of 550 ms following the development of the high at pin 28, due to the charging action of C7C1. When Q7A9 conducts, an Se scanner signal is applied to the PB1 decoder input. The voltage

236 Tuner and Frequency Synthesis Circuits

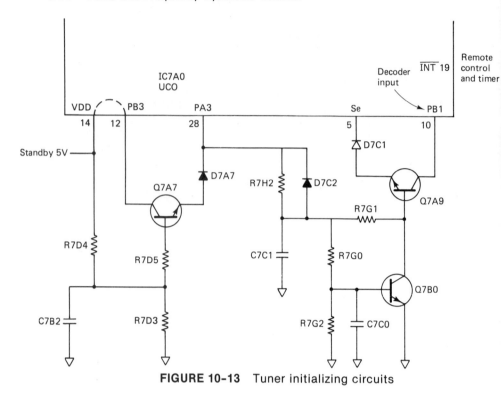

FIGURE 10-13 Tuner initializing circuits

across C7C1 is applied to Q7B0 through R7G0. The rise of the voltage at the base of Q7B0 is delayed an additional 20 ms by the charging action of C7C0. After 20 ms, Q7B0 turns on, disabling Q7A9 and removing the scanner signal. This completes the initializing process.

10-3 ANOTHER FREQUENCY SYNTHESIS TUNER SYSTEM

Figure 10-15 shows the FS tuner system of another VCR (that shown in Fig. 1-1c). Again, note that a single PLL microprocessor IC701 is used to control the FS tuning system. IC701 also performs other functions, such as supplying

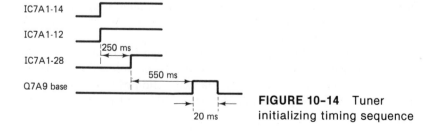

FIGURE 10-14 Tuner initializing timing sequence

Another Frequency Synthesis Tuner System 237

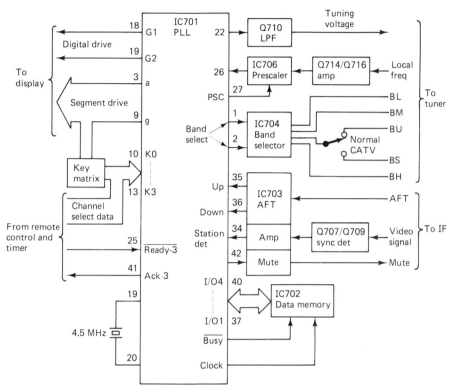

FIGURE 10-15 FS tuner PLL system block diagram

digit- and segment-drive signals for a two-digit channel display on the front panel. IC701 uses the segment drive to scan the keyboard and then monitors the selected signals at inputs K0 through K3 (pins 10–13). These signals contain channel data information. The K0 through K3 ports are also used as input ports to receive the channel-select data from the remote control microprocessor (Chapter 4), and the timer microprocessor (Chapter 14). *Handshaking lines* at pins 25 and 41 are used during transmission of channel-select data from the remote control and timer microprocessors, as described in Sec. 10-3.1.

The channel information programmed into memory in the tuning system is contained in IC702. The data bits are transferred between IC701 and IC702 via four I/O lines (pins 37–40). During the time that the PLL system is not properly tuned to a station, or if no station is on the air, the *mute output* (pin 42 of IC701) generates a low that is inverted to a high by IC703 and applied to the IF circuits (muting the audio, Chapter 11). For the PLL to recognize when the system is properly tuned to a station, IC701 monitors the *AFT and station-detect signals* at pins 34, 35, and 36.

IC701 generates two band-select code signals at pins 1 and 2. These signals are passed to band-selector decoder IC704, which (combined with the cable/normal switch) generates five band-switching signals: VHS low, VHF high, UHF, and CATV midband and superband. These band-switching signals are passed to the multiband tuner to select the proper band required for the channel requested (as discussed in Sec. 10-3.2).

10-3.1 Tuner PLL microprocessor communication

Figure 10-16 shows the communications arrangement among the timer, remote control, and tuner/PLL microprocessors. As shown, channel information from either the timer or remote microprocessors is passed to IC701 via a 4-bit parallel data bus. Communications are passed between the timer and remote microprocessors to establish which microprocessor is sending information. Upon receiving this communications signal, the other microprocessor goes into a standby or inhibit mode. The communications signal is sent via the transfer-1 and transfer-2 lines to the appropriate microprocessor.

As an example, if the timer microprocessor IC101 turns on in the timed-recording mode and sends channel-change information to IC701, IC101 pulls the transfer-1 line low, informing the remote-control microprocessor IC401 that IC101 is ready to send channel information, and that IC401 should go into standby.

The ready-3 and acknowledge-3 (ACK 3) handshaking lines are also used in this control data transfer between IC101 or IC401 and IC701. The ready-3 signal (pin 25 of IC701) informs IC701 that channel data bits are ready to be sent. IC701, when ready to receive the data, pulls the ACK-3 line high.

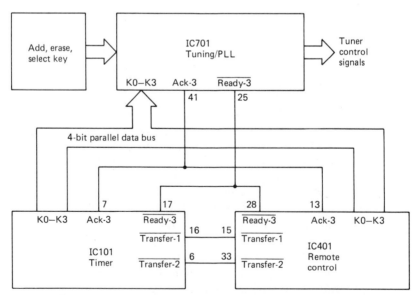

FIGURE 10-16 Communications arrangement among the timer, remote control, and tuner/PLL microprocessors

The following is a typical example of the communications sequence from the timer microprocessor IC101 to the tuner/PLL microprocessor IC701.

IC101 pulls ready-3 and transfer-1 lines low.

IC401 receives transfer-1 signal and goes into standby.

IC701 receives ready-3 signal and pulls the ACK-3 line high.

IC101 receives ACK-3 signal, outputs 4-bit data for the tens digit, returns ready-3 and transfer-1 line high.

IC701 receives data and returns ACK-3 line low.

Sequence repeats to send ones-digit data.

10-3.2 FS tuning system

Figure 10-17 shows the FS tuning system PLL circuits. As shown, the system is located on two boards: the tuning/PLL board and the tuner/demodulator board. The multiband tuner is controlled by circuits within IC701, which monitors various signals from the IF demodulator circuits to know when a station is being received. These signals are the AFT up/down (pins 35 and 36) and station detector signal (pin 34).

The detected video from the output of the IF demodulator circuits is passed to a sync amplifier and detector Q707-Q709. The detected sync signal (station detect) is passed to pin 9 of IC703, amplified, and applied to the station-detect input at pin 34 of IC701. When a station is properly tuned, the sync is detected from the video signal and applied as a high to pin 9 of IC703. This applies a high to pin 34 of IC701, indicating to IC701 that video with sync is present.

To maintain proper tuning, IC701 monitors the AFT up/down signals, which are supplied from the IF demodulator to pin 5 of IC703. The AFT circuit of IC703 is a *window detector,* monitoring the AFT voltage and outputting a high at pins 1 or 2, depending on the magnitude and direction the AFT voltage swings (should the tuner oscillator frequency drift).

When a channel has been selected, band-switching information is supplied from IC701 (at pins 1 and 2) to IC704, which develops four band-switching outputs. (One of these outputs is switched by the normal/cable switch located on the rear panel.) When the normal/cable switch is used, *five* possible band-switching signals are passed to the multiband tuner: VHF low, VHF high, UHF, midband, and superband.

The local oscillator within the multiband tuner passes a *sample carrier signal* through a shielded cable to the local oscillator amplifier Q714-Q716. The amplified local oscillator signal is then passed to pin 2 of the prescaler IC706. The output from IC706 (pin 5) is then passed to the sample input, pin 26 of IC701. When a channel has been selected, circuits within IC701 produce

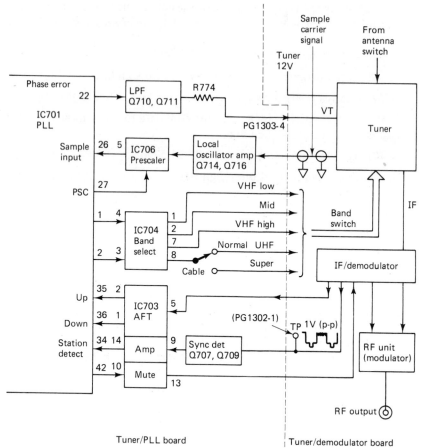

FIGURE 10-17 FS tuning-system PLL circuits

the appropriate number of pulse swallow control (PSC) pulses at pin 27. The PSC pulses are routed to the PSC control input of IC706. The relationship between IC701 and IC706 forms the PSC PLL functions, essentially as described in Sec. 10-1.1.

The divided-down local oscillator signal at pin 26 of IC701 is divided down again within IC701 and compared to an internal 5-kHz reference signal. The phase error of these two signals appears at pin 22 of IC701 and is applied to the low-pass filter Q710–Q711. The dc output from the low-pass filter is applied to the multiband tuner through R774. This voltage sets the tuner control as necessary to get the proper local oscillator frequency for the channel selected.

Another Frequency Synthesis Tuner System

FS tuning system troubleshooting/repair. The most common symptoms for failure of the FS tuning system are a combination of *no stations received, picture snowing,* and *audio noisy.* The first troubleshooting step is to isolate the problem to the tuner/IF demodulator/RF modulator (on the tuner/demodulator board, in this case) or the PLL system (on the tuning/PLL board in this VCR). Start by checking for power to all ICs and components. For example, the tuner requires +12 V in the circuit of Fig. 10–17. Once you are satisfied that power is available to all components, start the isolation process.

Select a local channel in the area and confirm that the *band-switching signal* for that particular channel appears at the tuner input and band-select output. For example, if a local station exists (and is on the air) between channel 2 and 6 (say, channel 4), select channel 4 and check that the VHF-low band-switching signal is high. This signal should appear at pin 1 of IC704 and at the tuner band-switch input (through components on the tuner/demodulator board). If the band-switching signal is not at pin 1 of IC704, suspect IC704.

Next, apply a *substitute tuner control voltage* at the input of the tuner/demodulator board (at PG1303-4). This substitute tuning control voltage can be obtained from an external variable power supply (0 to 30 V), or from a circuit such as shown in Fig. 10–18 (as recommended in the service literature).

Using an oscilloscope, monitor the *video signal* at the video test point (PG1302-1). (This signal is the usual composite video IF output, with both video and sync information, at about 1 V peak-to-peak.)

If a video signal appears at the test point, but there is no picture on the TV screen, suspect the RF modulator. (On those VCRs which have a video output to a monitor-type TV, it is relatively easy to isolate the problem to either the IF demodulator or RF modulator.)

If the video signal does not appear, suspect a problem in the complete tuner/demodulator assembly (multiband tuner, IF demodulator, RF antenna switch). In most VCRs, this means replacing the complete assembly or possibly replacing the individual tuner and IF demodulator ICs.

If you can tune in stations, using the substitute tuner control voltage, the problem is most likely in the tuner/PLL board components, rather than the tuner/demodulator components.

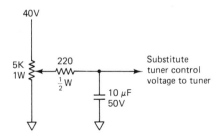

FIGURE 10-18 Tuning voltage-substitution circuit

242 Tuner and Frequency Synthesis Circuits

IC701 PLL output		Band designation	Channel	IC704 output Hi
BO-1 Pin 2	BO-2 Pin 1			
Low	Low	Low VHF 2 ↔ 6 —————— Midband A–5	2 ↔ 6 —————— 69	Pin 1
High	Low	Midband A–4 ↔ A–1 A ↔ I	70 ↔ 73 14 ↔ 22	Pin 2
Low	High	High VHF 7 ↔ 13 —————— Low superband J ↔ O	7 ↔ 13 —————— 23 ↔ 28	Pin 7
High	High	High superband P ↔ W —————— Hyperband W+1 ↔ W+26 —————— UHF 14 ↔ 83	29 ↔ 36 —————— 37 ↔ 62 —————— 14 ↔ 83	Pin 8

FIGURE 10-19 Band-switching decoder truth table

Tuner/PLL components troubleshooting repair. If video appears at test point PG1302-1 (and a picture appears on the TV screen), using a substitute tuner control voltage, check for a high at the station-detect input of IC701 at pin 34. If the station-detect input is missing, suspect IC703 or the sync detector circuits Q707–Q709.

If the station detect signal at IC701-34 is normal, monitor the AFT up/down inputs to IC701, pins 35 and 36, while changing the substitute tuning voltage. *If there are no logic changes* at pins 35/36 as you tune through a station, suspect the window detector circuit within IC703 (or possibly the AFT detector within the IF demodulator).

If the inputs at pins 35 and 36 appear to be normal, check for a sample local-oscillator signal at pin 26 of IC701. If the local-oscillator signal is missing, suspect IC706, the local-oscillator amplifier Q714–Q716, or possibly the shielded cable from the tuner. If the input at pin 26 of IC701 appears to be normal, but the outputs at pins 1 and 2 are absent or abnormal, suspect IC701.

Figure 10-19 shows the logic at pins 1 and 2 of IC701 and at the band-switching inputs of the tuner, for all the channels.

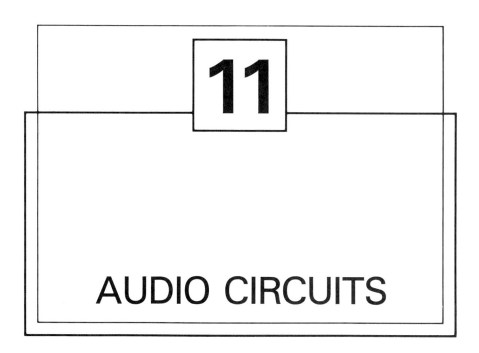

AUDIO CIRCUITS

This chapter is devoted to troubleshooting and repair of VCR audio circuits. A significant feature of most modern VCRs is the ability to record and play back audio signals in *stereo*. When playing back a stereo tape, the right and left audio channels are usually available at right and left audio jacks (often located on the rear panel), in addition to audio modulation of the RF unit. An external stereo amplifier may then be used to amplify the two audio channels for application to a stereo speaker system. (Note that stereo reproduction usually requires that the VCR be interfaced with a stereo amplifier capable of accepting *audio line-out signals*.)

To accommodate stereo capabilities, the conventional audio head found on early-model VCRs (typically with a 1-mm track width) is replaced by *two audio heads* mounted on the same audio/control (A/C) head assembly. Each audio head produces a significantly narrower individual track when compared to the track produced by a single audio head. These narrower audio tracks result in lower signal-to-noise (S/N) ratio, and generally require some form of compensation. Most modern VCRs compensate for the lower S/N ratio with the Dolby noise reduction system, which increases the S/N ratio to an acceptable level.

Most modern VCRs have a form of *audio monitor select* switch which permits the user to select *right and left* channels (concurrently) for normal stereo operation, or to select *right or left* channels (individually) for two-channel independent operation (such as bilingual operation). However, if a stereo tape is played and the output is monitored via the RF modulator, the right and left channels are electronically *added,* producing a normal monaural audio signal. So there is no loss in normal sound quality when monitoring a stereo tape on a conventional (monaural) TV set.

As in the case of early-model VCRs, most modern VCRs include an *audio dub* feature. However, with stereo VCRs, the audio dub feature is used on only one track (typically the right track). The other track is not affected. This permits the user to record *sound on sound*.

On some modern VCRs, it is possible to use the remote to control the audio output at the rear panel. However, on most VCRs, remote control of the volume level is available only when the RF modulator output is used.

Before we go into audio circuits, let us review the Dolby system (as it applies to VCRs) and stereo audio-head construction.

11-1 STEREO AUDIO-HEAD CONSTRUCTION

The VHS standard permits an audio-track width of 1 mm, recorded on the very top edge of the video tape. Figure 11-1 shows the relationship between the monaural/stereo audio track and the video tracks. Figure 11-2 shows the

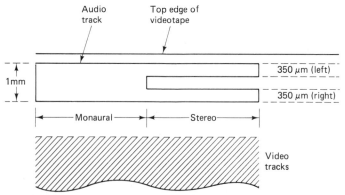

FIGURE 11-1 Relationship between the monaural/stereo audio track and the video tracks

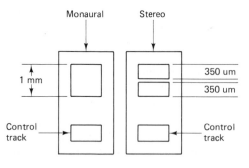

FIGURE 11-2 Relationship between monaural and stereo heads

relationship between monaural and stereo heads. To keep the stereo heads within the allowable 1-mm width, the stereo heads are reduced to 0.35 mm each. This size reduction allows a spacing or *guard band* of 0.3 mm between the audio heads, providing good separation of the two audio signals.

11-2 THE DOLBY SYSTEM

Background or *hysteresis noise* caused by the physical construction of videotape is a fact of life in any audio tape-recording process. In VCR recording, this hysteresis or background noise occurs in the frequency range of about 5 to 10 kHz. In VCRs with monaural audio, the background noise is normally not noticeable due to the level of the audio signals present. However, in stereo VCRs where the head size (and audio signal strength) is greatly reduced from that of monaural, background noise can become as great (or possibly greater) than the audio signal, thereby covering the audio signal with noise.

The Dolby noise-reduction system processes the audio signal to reduce the effects of the hysteresis/background noise *by changing the dynamic range characteristics* of the audio signal. There are several types of Dolby systems presently available, ranging from very expensive commercial studio systems to consumer-type systems. Most VCRs use what is known as the Dolby-B system, which provides Dolby noise reduction in the frequency range of hysteresis/background noise (5 to 10 kHz).

Troubleshooting Dolby systems. Keep the following points in mind when troubleshooting any audio circuits with Dolby noise reduction. Most (if not all) of the Dolby circuits are contained within a single IC. Typically, there are encoder circuits (for record) and decoder circuits (for playback). You cannot get at individual Dolby circuits. Instead, you must replace the IC (as is the case with all IC equipment). However, before you tear into the audio circuits with soldering tool and pickax, consider the following.

You get the best audio performance (stereo or monaural) when recordings that were made with the Dolby circuit on are played back with Dolby on. Equally true, recordings made without Dolby give the best performance when played back with the Dolby off. Playing back a non-Dolby recorded signal through a Dolby processing system causes the high-level signals of the audio information to be larger or smaller than normal. Conversely, playing back a tape recorded in Dolby with the Dolby turned off causes the high-frequency audio to be decreased in amplitude. Either way, you get distortion.

So, if you are troubleshooting an audio distortion problem on a VCR with Dolby, try switching the Dolby circuit on and off. If audio distortion is cleared in one position of the Dolby on/off control, you have isolated the problem to one of "probable operator trouble" rather than a circuit malfunction. Of course, if you record with Dolby on and play back with Dolby,

246   Audio Circuits

but get distortion, the problem is real. In that case, try recording without Dolby and playing back without Dolby. If the distortion disappears, you have localized the malfunction to the Dolby circuit IC or the control lines that switch Dolby in or out. If you get distortion where both record and playback are without Dolby, you may as well read the rest of this chapter.

11-3 TYPICAL AUDIO RECORD CIRCUITS

Figure 11-3 shows the audio record circuits of a typical VCR (that shown in Fig. 1-1a). Note that most circuit components, except those involved in switching, are contained within ICs. Also note that only the left audio channel

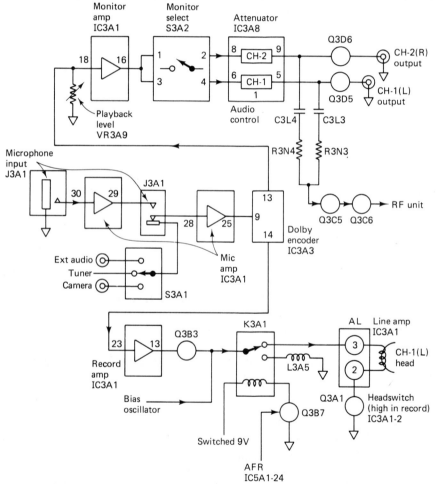

FIGURE 11-3 Typical audio record and monitoring circuits

is shown. The right audio channel is identical, with the exception of component nomenclature and relay K3A1 (which is not used in the right audio-channel circuit).

The source of audio to be recorded is selected by the three-position signal selector switch S3A1. In the *tuner* position, normal monaural audio from the tuner/IF circuits is directed to *both right and left* audio channels simultaneously. In the *external* position, audio signals entering the external *right and left* audio input jacks are directed to the appropriate audio circuit. In the *camera* position, audio from the camera microphone is applied to the audio record circuits.

Audio from the signal selector switch S3A1 is applied to the input of the line amplifier, at pin 28 of IC3A1, via the microphone input jack J3A1. If an auxiliary microphone (not the camera microphone) is inserted into J3A1, audio from S3A1 is automatically disconnected from line amplifier IC3A1, and the auxiliary microphone output is applied instead. So the microphone input has priority *over all other audio inputs* in the record mode.

The left-channel source audio is amplified by the line amplifier within IC3A1 and appears as an output at pin 25. This amplified output is applied to the Dolby encoder IC3A3 at pin 9. The audio (with Dolby encoding) is taken from IC3A3 at pin 14 and applied to the record amplifier within IC3A1 at pin 23. The output of the record amplifier, at pin 13 of IC3A1, is added in Q3B3 to the bias signal from a *bias oscillator*. The combined audio and bias signals are then directed to the left-channel audio head via relay K3A1.

During record, system control generates a high head-switching command at pin 2 of IC3A1. This command turns on Q3A1 and provides the audio head with a ground-return path so that the combined audio and bias can be recorded by the head.

During audio dub, when audio is being recorded on the right audio channel, relay K3A1 disconnects the left-channel audio head, so the left-channel audio channel is not affected by the audio dub operation. When audio dub is selected, system control makes the AFR line high, turning on Q3B7 and energizing K3A1. This disconnects both the bias oscillator and record amplifier from the left-channel audio head input, and the left audio track remains intact (not erased, no new material added). Since the right channel does not contain relay K3A1, the right audio track is erased and new material is recorded in audio dub.

11-4 TYPICAL AUDIO MONITORING CIRCUITS

Figure 11-3 also shows the audio monitoring circuits. As with the audio record circuits (Sec. 11-3), the right- and left-channel circuits are identical except for component nomenclature.

A sample of the left-channel audio signal is output from pin 13 of the

Dolby encoder IC3A3 for monitoring purposes. Note that the signal at pin 13 *does not have Dolby correction*. The monitor signal is applied to a monitor amplifier within IC3A1 at pin 18. The amplified output of the monitor amplifier is applied to pins 1 and 3 of IC3A5 (which is discussed further in Sec. 11-6).

IC3A5 contains several analog switches, controlled by the front-panel monitor-select switch S3A2, allowing the user to select three types of audio monitoring:

1. Left and right channels—the normal stereo position, with right- and left-channel audio signals output from the respective output jacks.
2. Right channel only—right-channel audio is output from *both* the right and left output jacks simultaneously.
3. Left channel only—left channel audio is output from both output jacks simultaneously.

The audio signal from IC3A5 (right-channel audio from pin 2 and left-channel audio from pin 4) is directed to IC3A8, pins 8 and 6, respectively. IC3A8 contains right- and left-channel attenuation amplifiers, the gain of which is controlled by a dc voltage at pin 1. The higher the voltage at pin 1, the more attenuation applied to the audio signal.

Since the voltage at pin 1 is variable (controlled by the remote unit, Chapter 4), the VCR may be used as a remote for the monitor TV receiver, with full remote control of the sound level. Also, since sound-level control occurs prior to the audio jacks on the VCR, remote volume control is available when using either the direct audio output jacks or the RF output, or both.

Right-channel audio is output at pin 9 of IC3A8 and is applied to the right audio output through Q3D6. The left-channel audio output, from pin 5, is directed to the left audio output jack through Q3D5. Samples of right and left audio signals are added to form a monaural audio signal for the RF modulator through Q3C5 and Q3C6.

11-5 TYPICAL PLAYBACK AUDIO CIRCUITS

Figure 11-4 shows the audio playback circuit of a typical VCR (that shown in Fig. 1-1a). Again, note that most circuit components, except those involved in switching, are contained within ICs. Also, only the left audio channel is shown. The right channel is identical except for nomenclature.

Audio from the left audio head is output from pin 2 of the AL connector. Pin 3 of the AL connector is the audio-head ground return and is connected to ground through Q3A3 and Q3A5. In playback, Q3A3/Q3A5 are turned on by a low on the SRV/REC line applied to Q3A7/Q3A9/Q3B1 (providing a ground return to the head).

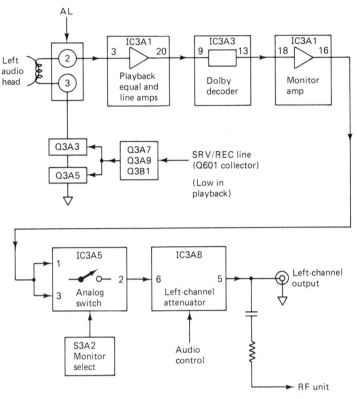

FIGURE 11-4 Typical audio playback circuits

Audio from pin 2 of the AL connector is amplified by the playback equalizer and line amplifiers within IC3A1. The amplified left-channel audio is output at pin 20 of IC3A1 and is applied to pin 9 of the Dolby decoder in IC3A3. Audio is output at pin 13 of IC3A3.

The remaining audio playback path is identical to the monitor path described in Sec. 11-4. That is, left-channel audio is directed to the left audio output jack and the RF modulator via the monitor amplifier in IC3A1, the analog switches in IC3A5, and the left-channel attenuator in IC3A8.

11-6 TYPICAL MONITOR SELECT SWITCH

As discussed, the user has a choice of selecting right-channel audio only, left-channel audio only, or right- and left-channel audio (stereo) on many modern VCRs. This selection is done by a monitor select switch circuit, such as that shown in Fig. 11-5.

When the monitor select switch S3A2 is in the center position (R-CH +

250   Audio Circuits

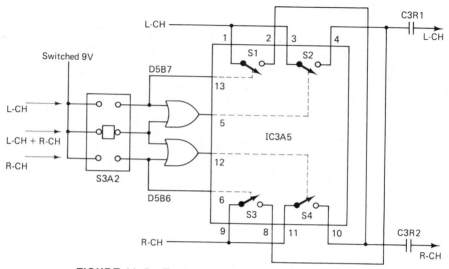

FIGURE 11-5 Typical monitor select switch circuit

L-CH), right-channel audio is output from the right audio output, and left-channel audio is output from the left audio output jack. So if a stereo tape is played, a stereo amplifier may be connected to the audio output jacks, and recorded stereo is reproduced.

With S3A2 in R-CH + L-CH, switched 9 V is directed to one input of each OR gate D5B7/D5B6. The OR gate outputs drive pins 5 and 12 of IC3A5 high, closing analog switches S2 and S4. In this condition, left-channel audio is directed to the left-channel output through S2, and right-channel audio is directed to the right-channel output through S4.

When S3A2 is in L-CH, pins 5 and 13 of IC3A5 are driven high, closing S1 and S2. Left-channel audio is applied to the left output through S2 and to the right output through S1. Analog switches S3 and S4 are open, so no right-channel audio passes.

When S3A2 is in R-CH, pins 6 and 12 of IC3A5 are driven high, closing S3 and S4. Right-channel audio is applied to the right output through S4 and to the left output through S3. Analog switches S1 and S2 are open, so no left-channel audio passes.

11-7 TYPICAL AUDIO CROSSTALK CANCELLER CIRCUITS

Some VCRs with stereo capability include a crosstalk canceller circuit such as shown in Fig. 11-6. The crosstalk canceller eliminates the leak of magnetic flux from the right-channel audio record/playback head to the left-channel head during monaural audio dub (when only the right channel is being dubbed,

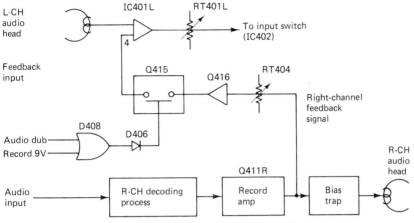

FIGURE 11-6 Typical audio crosstalk-canceller circuits

and the left channel is supposed to remain intact). Such leakage can cause crosstalk. (Typically, some of the dubbed information can get into the left channel.)

In the circuit of Fig. 11-6, the output (obtained from between the record amplifier and bias trap) of the right channel is applied to the feedback input of a preamplifier (IC401L, pin 4) of the left-channel audio system. This right-channel feedback signal is applied through RT404, Q416, and Q415 (which is turned on in either audio dub or record modes).

The feedback signal is adjusted (by RT404) to the *same level* as any crosstalk signal (generated by leaking magnetic flux) that may appear at the non-feedback input of IC401L (from the left-channel head). Since the two signals (crosstalk and feedback) are of the same amplitude but are applied to opposite inputs, the signals are cancelled in IC401L, eliminating any crosstalk on the left channel. Note that *when properly adjusted,* the feedback signal has no effect on the normal left-channel audio. So if you are troubleshooting a crosstalk problem during or after audio dub operation, always check any crosstalk adjustments first.

11-8 AUDIO CIRCUIT TROUBLESHOOTING AND REPAIR

The audio circuits for most VCRs are straightforward direct paths of input to output (from source to audio head during record, and from head to output or RF modulator during playback). Troubleshooting for the audio system involves basic signal tracing with audio-monitor devices such as meters and oscilloscopes. For that reason, we do not go into audio circuit troubleshooting in any detail. Instead, here are some tips for audio circuit troubleshooting.

If you are troubleshooting a stereo or two-channel VCR, *compare the readings* between the channels. The audio level readings should be essentially the same for both channels.

Always *check any adjustments* in the audio system first. For example, the playback-level adjustment VR3A9 in Fig. 11-3 sets the audio playback level for both the RF modulator and monitor outputs.

Check the frequency and level of the *bias signal.* Usually the audio bias is taken from the same oscillator that provides an erase signal to the full-erase head. The bias oscillator frequency (typically in the 60- to 70-kHz range) is usually not critical, but the amplitude of the bias signal is important. (A low level of bias can fail to erase previous recordings.)

It is usually difficult to measure the direct output of an audio head. However, the head output is usually fed to a line amplifier and/or playback equalization amplifier (such as shown in Fig. 11-4). The service literature usually gives some hint as to the audio signal level at the amplifier output. Of course, if you are dealing with two audio heads (on the same stack) and both heads show approximately the same signal level, it is reasonable to assume that the heads are good—but do not count on it.

Look for any switches, jacks, relays, and so on, that *may interrupt the audio signal path.* Most VCR audio systems have many such devices in the audio circuits. For example, as shown in Fig. 11-3, the contacts of microphone plug J3A1 between pins 28 and 29 of IC3A1 connect the microphone (and disconnect all other audio sources) when an auxiliary microphone is plugged in. If the contacts are poor (dirty, bent, etc.), with no microphone in place, the audio signal can be attenuated (or possibly interrupted). The same applies to the contacts of relay K3A1, which connect the combined left-channel audio record/bias signals to the audio head (except during audio dub).

Look for any special head-switching circuits. An example is the ground-return transistor Q3A1 (Fig. 11-3) for the left-channel audio head. Check that the head-switching command (a high from system control, in this case) is available, and that the transistor is turned on.

Keep in mind that the audio heads must be cleaned, as is the case with video, control track, and full-erase heads. Refer to Sec. 2-4 for cleaning procedures.

Finally, before you decide that an audio circuit IC is defective and must be pulled, always check all connections to the IC, especially the B+ or other power connections.

12

VIDEO CIRCUITS

This chapter is devoted to troubleshooting and repair of VCR video circuits. Although we cover both luminance and chroma sections of the VCR, we do not dwell on video circuits. There are two reasons for this approach.

First, although the luminance circuits of any VCR are very complex, they are not the major cause of trouble. Mechanical problems are on top of the list, closely followed by servo and system control troubles.

Second, although many circuits are involved, most of the video circuits are usually found in three or four ICs (in most modern VCRs). If all else fails, you can replace the few ICs, one at a time, until the problem is solved. (If only mechanical problems were that simple!) The one major exception to this applies to the adjustment controls in video circuits. The adjustment controls are found outside the ICs. However, when you go through the adjustment procedures recommended in the service literature (as discussed at the end of this chapter), you simultaneously localize faults in the adjustment control circuits.

The video circuits described in Chapter 1 generally apply to most modern VCRs. However, since we concentrate on VHS circuit details in Chapter 1 (since VHS seems to dominate the VCR market), we provide the same detailed coverage for Beta video circuits here (since Beta is still alive and well).

Again, keep in mind that each VCR (Beta or VHS) has its own unique set of circuits which you must check out during troubleshooting (using the service literature, hopefully). However, to give you a head start in video-circuit troubleshooting, we conclude this chapter with overall luminance and video troubleshooting tips that apply to all VCR video circuits.

12-1 INTRODUCTION TO BETA CIRCUITS

The majority of the circuits described here are part of the Sanyo VCR system. These VCRs, described as "Betacord" or "Betavision," use the conventional Beta color-under recording described in Sec. 1-4, and the Beta high-density PI (phase inversion) described in Sec. 1-5. The VCR described here has two tape formats, called BII (Beta 2) and BIII (Beta 3). The VCR uses a BII/BIII

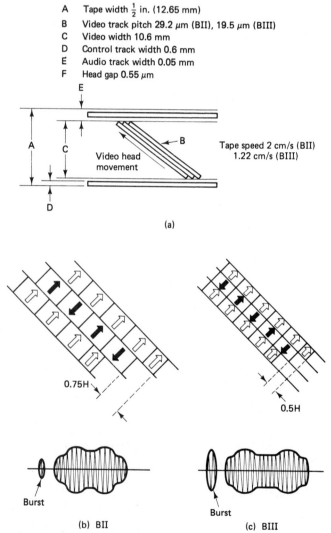

A   Tape width $\frac{1}{2}$ in. (12.65 mm)
B   Video track pitch 29.2 μm (BII), 19.5 μm (BIII)
C   Video width 10.6 mm
D   Control track width 0.6 mm
E   Audio track width 0.05 mm
F   Head gap 0.55 μm

FIGURE 12-1 Major differences between the BII and BIII systems

switching method which permits recording of the conventional BII system, in addition to the extremely high-density BIII recording system.

Figure 12-1 shows the major differences between BII and BIII. As shown in Fig. 12-1a, the tape formats are essentially the same for BII and BIII, except in video-track pitch and tape speed. (The slower tape speed permits longer playing time for a given amount of tape.) However, strictly speaking, the BIII system records on a true zero guard band, while the BII system records with a guard band of about 2 μm.

Another difference between BII and BIII is alignment in the color PI system. The Beta system uses 1H phase reversal of the chroma signal (recorded at 688 kHz). As shown in Fig. 12-1c, the BIII system uses a 0.5H alignment of phase reversal, rather than the 0.75H alignment for BII (shown in Figs. 12-1b and 1-12).

When BIII is selected, the burst of the recording 688-kHz chroma signal is amplified by 6 dB. This makes the recording current of the burst about the same level as a signal with high color saturation and improves the signal-to-noise (S/N) ratio. By improving the S/N ratio of the color burst signal, the jitter cancel effect of the APC circuit at the time of playback is improved considerably. This improvement is required since the recording density of BIII is much greater than for BII, and the effects of jitter are more noticed. After the BIII playback color burst has been used by the APC circuits, the amplitude of the burst is restored to normal by a burst deemphasis circuit.

Figure 12-2 is an overall block diagram of the Beta video circuits. As shown, the video signal system uses four ICs to process both the luminance (Y) and chroma (C) signal during record and playback.

12-2 BETA LUMINANCE CIRCUITS

As shown in Fig. 12-2, the Beta luminance signals are processed mainly by ICs Q1 and Q2. IC Q1 has both recording and playback functions, as shown in the simplified diagrams of Figs. 12-3 and 12-4. IC Q2 is used primarily in playback, as shown in Fig. 12-5.

During record, the video signal from the tuner (Chapter 10) is applied through the record/playback changeover switch in Q1 to the AGC (automatic gain control) system. Note that the AGC system consists of SYNC AGC in Q1 and the PEAK AGC (Q10-Q14) outside the IC.

12-2.1 AGC circuit and Y-FM modulator

Two AGC systems are used to accommodate video signals with different sync/signal ratios. In the case of an input signal with a sync/signal ratio of about 30% or more, the SYNC AGC provides a fixed sync signal level. When the sync/signal ratio is less than 30%, the AGC output can become excessive with

256 Video Circuits

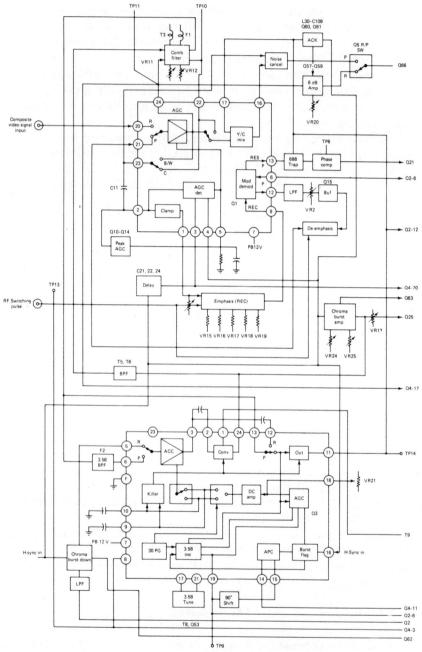

FIGURE 12-2 Overall block diagram of Beta video circuits

Beta Luminance Circuits 257

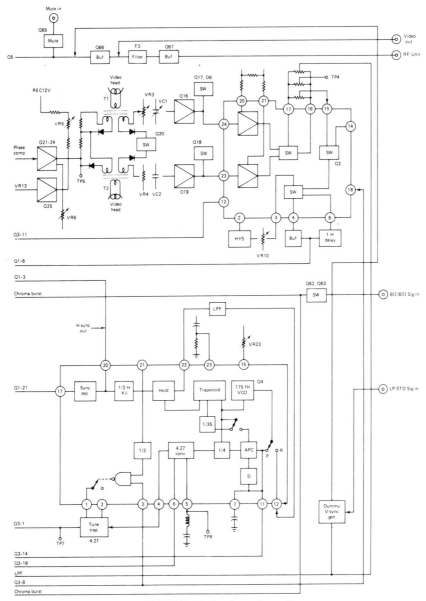

FIGURE 12-2 (continued)

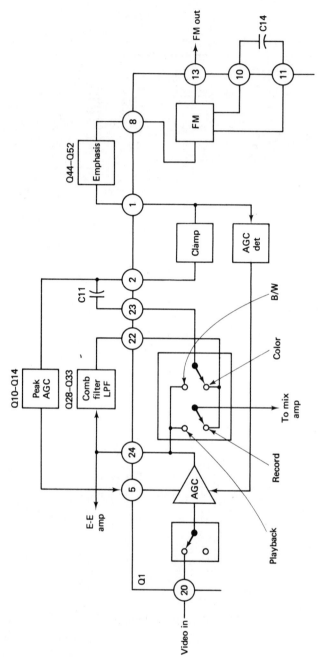

FIGURE 12-3 Typical Beta luminance-signal circuit

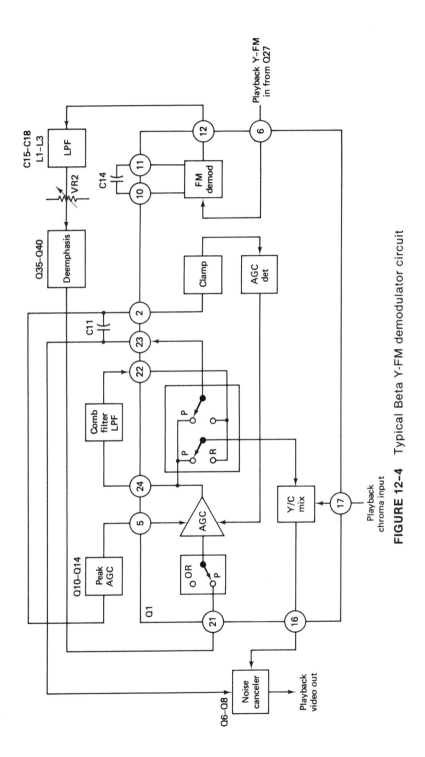

FIGURE 12-4 Typical Beta Y-FM demodulator circuit

259

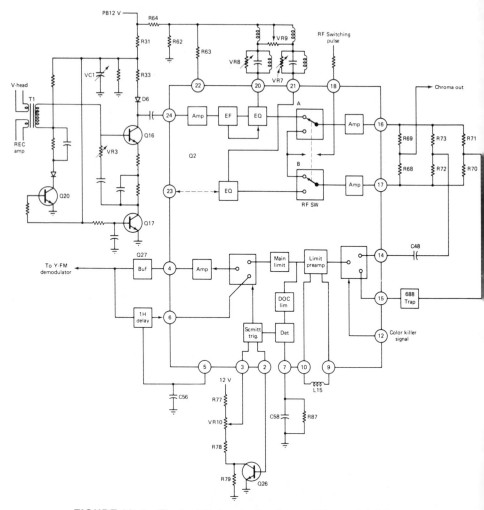

FIGURE 12-5 Typical Beta playback amplifier and DOC circuit

only the SYNC AGC. This causes overmodulation of the FM modulator and RF unit, resulting in buzz, and so on. When the sync/signal ratio is low, the PEAK AGC circuit takes over to prevent an excessive AGC output. With either circuit, the AGC signal output is applied to the comb filter circuit through pin 24 of Q1.

When the input signal is color, the signal from pin 24 passes the comb filter and low-pass filter. The luminance signal (with color removed) is returned to Q1 at pin 22. When the input is black and white, the signal is applied directly to pin 23 by a switch in Q1. Either way, the output from pin 23 is applied through C11 and pin 2 to a sync-tip clamp within Q1. The output of the sync-tip clamp is applied to an external emphasis network through pin 1. The emphasis network contains preemphasis, white-clip and black-clip circuits, as well as several adjustments. VR15 is used for sync-tip modulation frequency adjustment. VR17 is used for frequency shift adjustment, whereas VR16 controls the half-H adjustment. The white-clip and black-clip circuits are adjusted by VR18 and VR19, respectively. The output of the emphasis network is applied to the FM modulator.

The output of the FM modulator is applied through a 688-kHz trap and a phase compensation circuit to the Y record amplifier Q21. The trap is controlled by a color-killer signal from pin 17 of Q1. When the video signal is black and white, the trap removes the 688-kHz color signal.

12-2.2 Record amplifier

The recording amplifier for the Y-FM signal consists of Q21–Q24. The Y-FM signal (at the specified recording current) is applied to the video heads through T1 and T2. VR6 is used for adjustment of the recording current value. VR5 is used to adjust the frequency characteristic of the recording current. Q25 is the recording amplifier for the color signal (after being converted to 688 kHz). This color signal is superimposed on the Y-FM signal (at the specified level), and both signal currents are supplied to the video heads.

12-2.3 Playback amplifier and DOC

As shown in Fig. 12-5, the low-amplitude playback signal from the video heads passes step-up transformers T1 and T2 and is fed to the first-stage amplifiers Q16 and Q19 (which operate as low-noise amplifiers). The outputs of Q16 and Q19 are applied to cascade amplifiers within Q2. (Note that T1 and T2 operate as input/output transformers, supplying recording current to the video heads during record and stepping up the low-amplitude playback signals during playback.)

The frequency characteristics of the playback signals are adjusted by VC1, VR3, and VR8 for channel A, and VC2, VR4, and VR7 for channel B. The balance between channels A and B is adjusted by VR9. The frequency-compensated or adjusted-playback signal is switched by the RF switching pulse (Chapter 6) at pin 18, and appears as an output from Q2 at pins 16 and 17. The chroma signal is taken from this output at the junction of R68 and R69, and is applied to the chroma processing circuits (Sec. 12-3).

One Y signal is taken from the output at the junction of R72 and R73 and is applied through C48 to pin 14 of Q2. Another Y signal is taken from the junction of R70 and R71, and is applied through a 688-kHz trap (C50–L11) to pin 15 of Q2. Both Y signals are applied to a color/black-and-white switch within Q2. This switch is operated by a color killer signal applied at pin 12. When the playback contains color, the signal at pin 14 is used in the normal manner. In the absence of color, the switch selects the signal from pin 15.

The output from the color/black-and-white switch is applied to the DOC circuit. The limiter suppresses any waveform distortion and passes the video signal to the detector and adjustable Schmitt trigger. VR10 sets the firing point of the Schmitt trigger and thus sets the point at which dropout compensation occurs. If dropout is present, the Schmitt trigger switches the signal so that the preceding 1H delayed signal (passed through the 1H delay line, external to Q2) appears at the output. The compensated signal at pin 4 is applied through emitter-follower buffer Q27 to the Y-FM demodulation circuit.

12-2.4 Y-FM demodulation circuit

The circuits of Q1, used at the time of recording, operate as a demodulator at the time of playback. This is shown in Fig. 12-4.

The signal from the playback amplifier and DOC circuit is applied to pin 6 of Q1 during playback. This signal is demodulated by circuits within Q1 and is applied through a low-pass filter (C15-C18, L1-L3) to a deemphasis circuit (Q35-Q40). This deemphasis circuit has the reverse frequency characteristics from the preemphasis at the time of recording. The deemphasis circuit is set for optimum characteristics by VR2.

After passing through the deemphasis circuit to obtain the correct frequency characteristics, the playback Y signal is returned to Q1 at pin 21 and is applied to the same AGC circuit used during record (Fig. 12-3). Reprocessing by the AGC (at playback) helps restore the playback signal to the same level that exists at the time of record. After passing the AGC, the Y signal is mixed with the chroma signal (at pin 17) and appears as the playback composite video signal at pin 16.

The Y signal is also applied through pin 23 to a noise-canceller circuit (Q6-Q8). The noise-canceller output is combined with the composite video signal at pin 16. In the noise canceller, the high-frequency noise component is removed, the phase is inverted, and with reverse phase and the same amplitude, the processed video signal is added back to the composite video signal, thus cancelling the noise.

12-2.5 Record/Playback and video output circuit

The playback composite video signal from pin 16 of Q1 is applied to the RF unit (and to a VIDEO OUT connector) through a record/playback switching system and various video output circuits, as shown in Fig. 12-2. Operation of these circuits is controlled by an IC analog switch Q5, as shown in Fig. 12-6. IC Q5 provides the necessary changeover for the E-E and playback signals. (As discussed in Chapter 1, when in the record mode, the record output is

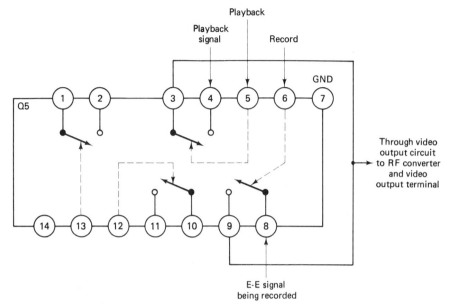

FIGURE 12-6 Record/playback switching system

connected to the playback input so that the video signal can be monitored on a TV set.)

As shown in Fig. 12-6, operation of switches within Q5 is controlled by signals at pins 5 and 6. During record, a signal is applied to pin 6, closing the switch across pins 8 and 9. This connects the E-E signal (signal being recorded) to the RF unit and VIDEO OUT terminals. During playback, a signal is applied to pin 5, closing the switch across pins 3 and 4. This connects the playback signals to the RF converter and VIDEO OUT.

As shown in Fig. 12-2, the video signal from pin 24 of Q1 is applied to a 6-dB amplifier (Q57–Q59) and is then applied to pin 8 of Q5 (Fig. 12-6). VR20 is used to adjust the E-E level from the 6-dB amplifier. The ACK (automatic color killer) circuit (Q60–Q61) functions to remove the burst signal from the E-E video if the signal is black and white, or if the color signal being recorded is extremely attenuated. The ACK does this burst removal by inserting a 3.58-MHz trap when a color-killer signal is applied. (This is the same color-killer signal applied to pin 12 of Q2, Fig. 12-5.)

After passing the noise-canceller circuit, the playback video signal is applied to pin 4 of Q5. Either the playback signal (pin 4) or the E-E signal (pin 8) is selected by operation of the switches in Q5. The output from Q5 (pins 3 and 9, tied together) is applied through the video output circuit to the RF unit and VIDEO OUT connector.

The video output circuit is composed of transistors Q65, Q66, Q67, and

264 Video Circuits

filter F3. Q65 operates as a muting circuit for the video signal. During playback, if there is no control signal output from the servo system because of a malfunction (or if the CTL signal is not properly recorded), the video signal is muted (cut off at Q65). This produces a black raster on the monitor TV screen. Transistor Q66 functions as an emitter follower, and provides a 75-Ω output impedance for the VIDEO OUT connector. Transistor Q67 operates as a buffer between F3 and the RF unit.

12-3 BETA CHROMA CIRCUITS

The chroma or color signals are processed mainly by ICs Q3 and Q4. The relationship of Q3 and Q4 to the remaining video circuits is shown in Fig. 12-2. Figures 12-7 and 12-8 show the Q3 and Q4 circuits in greater detail. Before discussing operation of these circuits, let us review the overall functions of the chroma circuits.

As shown in Fig. 12-9, the main function of the chroma circuits during record is to convert the 3.58-MHz color burst signal to a 688-kHz signal, and

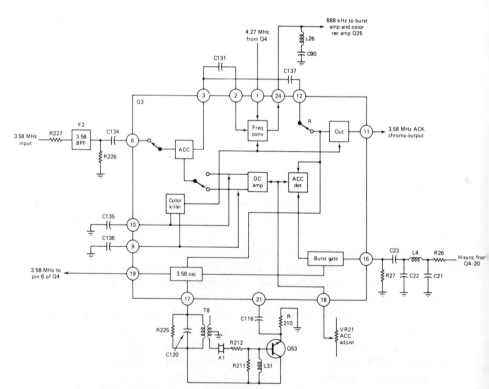

FIGURE 12-7 Typical Beta record-mode operation

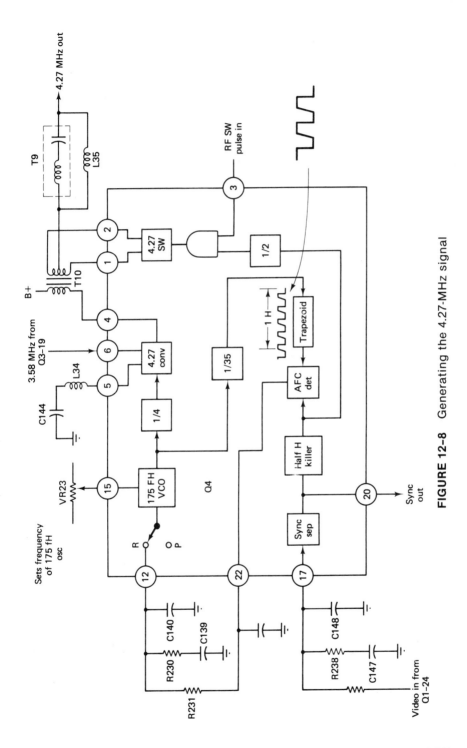

FIGURE 12-8 Generating the 4.27-MHz signal

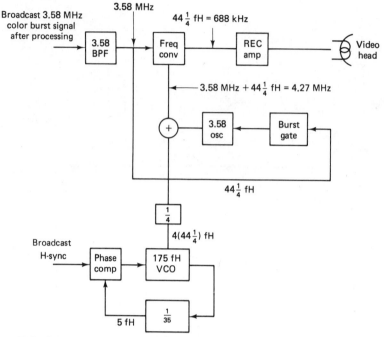

FIGURE 12-9 Basic functions of chroma circuits during record

to record this signal on tape. The 688-kHz signal is locked in phase and frequency to the broadcast 3.58-MHz and H-sync signals by AFT and APC circuits.

Because of zero guard-band recording, the 688-kHz signal is recorded with a phase reversal of 1H on every other track to remove crosstalk. As shown in Fig. 12-10, the function of the chroma circuits during playback is to convert the 688-kHz signal back to a 3.58-MHz color burst. This 3.58-MHz signal is locked in phase and frequency to the playback 688-kHz and H-sync signals. The 1H phase reversal is removed, and the 3.58-MHz signal is combined with the luminance signal to produce a composite video signal identical (hopefully) to the broadcast video signal.

12-3.1 Record mode

The composite video signal at the output of the AGC circuit within Q1 (Fig. 12-2) is taken from pin 24 and applied to pin 6 of Q3 through a comb filter and a 3.58-MHz bandpass filter F2. T3, VR11, and VR12 set the level of the video signal passing through the comb filter. Filter F2 passes only the 3.58-

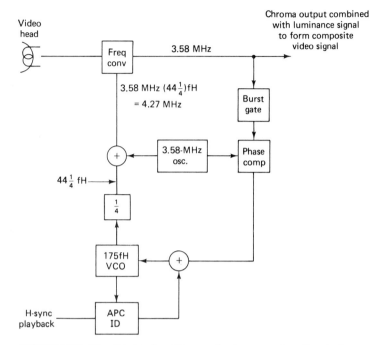

FIGURE 12-10 Basic functions of chroma circuits during playback

MHz signal applied to pin 6 of Q3 (Fig. 12-7). The H-sync is applied to the burst gate circuits within Q3 through pin 16.

The output of the 3.58-MHz oscillator, locked to the burst gate signal, appears at pin 19 of Q3. This signal is applied to a 4.27-MHz converter within Q4. Simultaneously, the 3.58-MHz signal at pin 6 of Q3 is applied to the frequency converter within Q3 through the ACC circuit and C131. The frequency converter also receives a 4.27-MHz signal from Q4 through pin 1 of Q3. The combination of the two signals at the frequency converter produces a 688-kHz signal which is applied (through pin 24 and a burst amplification circuit, as described in Sec. 12-3.2) to the color record amplifier Q25. Since the 4.27-MHz signal is phase-reversed at 1H on every other track, the 688-kHz color signal is recorded with the phase inversion.

During record, the 3.58-MHz signal at pin 3 of Q3 is applied through C137 and circuits within Q3 to appear as a 3.58-MHz ACK (automatic color killer) output at pin 11 of Q3. In the absence of a 3.58-MHz signal (black and white, or badly attenuated color) the ACK circuits function to cut off the color circuits (such as removal of the E-E video).

Q3 also contains ACC (automatic color control) circuits which maintain the level of the color signal. VR21 sets the level of the color signal at IV.

12-3.2 BIII burst amplification circuit

The 688-kHz signal to be recorded is amplified by 6 dB when the BIII mode is selected. This is done by the burst amplification circuit shown in Fig. 12-11. A burst gate pulse (H-sync) is applied to Q42, and a BII/BIII switching-control signal is applied to Q72 and Q73. The switching-control signal is high for BIII and low for BII. VR13 sets the record signal level during both BII and BIII.

Although the circuit is called a burst amplifier, the circuit is more like that of an attenuator. When BII is selected, Q73 is on and Q72 is off. During BII, the circuit appears as shown in Fig. 12-11b, where part of the 688-kHz

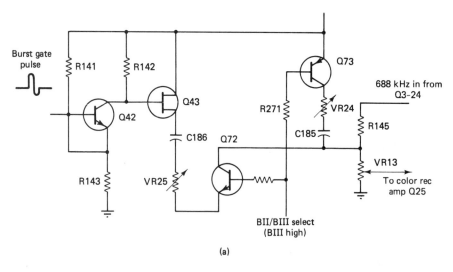

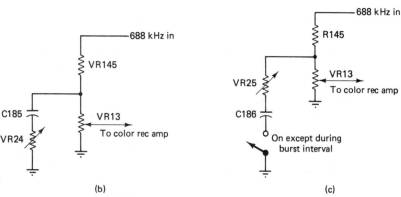

FIGURE 12-11 BIII burst-amplification circuit

signal to be recorded is passed through C185 and VR24 to ground. This attenuates part of the signal. The level of the recorded signal during BII is set by adjustment of VR24.

When BIII is selected, Q72 is on and Q73 is off (during the burst interval) and on at other times. During BIII, the circuit appears as shown in Fig. 12-11c, where part of the 688 kHz to be recorded is passed through VR25 and C186 to ground, at all times except during the burst period (when the burst gate is present). This attenuates the signal, except during the burst period. The net result is that the burst level is increased in relation to the remainder of the signal. VR25 sets the level of record during BIII.

12-3.3 Generation of 4.27-MHz signal during record

As shown in Fig. 12-8, the 4.27-MHz signal used in the color circuits is generated in Q4. ICQ4 receives a composite video signal (containing H-sync signals) at pin 17, RF switching pulses at pin 3, and a 3.58-MHz signal from Q3 at pin 6. The 4.27-MHz output is at pins 1 and 2. The H-sync pulses at pin 17 are separated by the sync separator circuits within Q4 and appear as an output at pin 20.

During record, the output of the 175fH VCO (voltage-controlled oscillator) is counted down by 35 to 5fH by circuits within Q4, and produces a trapezoidal wave. Simultaneously, the H-sync separated from video (at pin 17) is compared in phase with the trapezoidal wave, and the resultant error voltage applied to the 175fH VCO through pins 22 and 12 and a filter circuit (C139, C140, R230, R231). This filter smooths out the error voltage, which locks the 175fH VCR to H-sync. The exact frequency of the 175fH VCO is set by VR23.

The 175fH signal is divided by 4. The resulting 44¼fH signal (locked to the H-sync) is applied to a frequency converter within Q4. This frequency converter combines the 44¼fH with the 3.58-MHz signal from Q3 to produce a 4.27-MHz signal. Since the 44¼fH signal is locked to H-sync, and the 3.58-MHz signal is locked to the broadcast 3.58-MHz color signal, the 4.27-MHz signal from Q4 is thus locked to both signals. The 4.27-MHz signal is applied through T10 to the converter in Q3.

Transformer T10 acts as a carrier-phase inverter. The phase of the 4.27-MHz signals passing through T10 is reversed every 1H by operation of the circuits within Q4. These phase-inverter circuits receive both RF switching pulses and H-sync pulses, and function to alternately short pin 1 or pin 2 of Q4 to ground for each H-sync pulse. This shorts either end of the T10 secondary every 1H, but only during track A. The RF switching pulses override the H-sync pulses when track B is being traced by head B. Each time one end of the T10 secondary is grounded, the center-tap output is inverted in phase.

270   Video Circuits

12-3.4 Playback mode

The basic function of the video circuits during playback is shown in Fig. 12-10. Figure 12-12 shows operation of the Q3 circuit during playback. As shown, the 688-kHz output from the preamplifier in Q2 is applied to pin 5 of Q3 through a low-pass filter (C130, C133, L33) at the time of playback. When BIII is selected, the burst is amplified 6 dB by a separate circuit. During BIII playback, the burst signal is restored to a normal level by a burst deemphasis circuit.

In either BII or BIII modes, the 688-kHz signal at pin 5 of Q3 is passed through the ACC circuit. Note that the ACC circuit of Q3 has an independent dc amplifier circuit for each field to improve transient response. Fields A and B are switched by the RF switching pulses applied at pin 8. The 688-kHz signal from the ACC circuit is applied to a frequency converter through C131. The frequency converter also receives a 4.27-MHz signal generated in Q4 through pin 1. The resultant 3.58-MHz signal is applied through a bandpass filter (T5, T6) and a comb filter (Q28-Q30, DL2) to a chroma output amplifier in Q3. The comb filter restores the phase reversal introduced by Q4 and T10 (Fig. 12-8). The 3.58-MHz playback signal from the chroma output amplifier is applied through pin 11 and the ACK circuits to pin 17 of Q1, where the chroma signal is mixed with the luminance signal.

As shown in Fig. 12-12, the 3.58-MHz signal at pin 13 of Q3 is also applied to an APC (automatic phase control) circuit. This APC circuit also receives H-sync signals through a burst gate (pin 16) and 3.58-MHz signals from pin 19 (shifted 90° by C117, L32, and R213). The burst gate signals turn on the APC detector during the time of color burst. The 3.58-MHz crystal reference signal is compared with the phase of the 3.58-MHz playback signal. Any difference in phase produces an error signal at pin 14. This error signal is applied to pin 11 of Q4 to control the phase of the 175fH VCO.

12-3.5 Generation of 4.27-MHz signal
during playback

At the time of recording, Q4 produces 44¼fH by a one-fourth countdown of the 175fH VCO, locked to the H-sync of the video signal. Similarly, the 4.27-MHz signal is produced from the 3.58-MHz signal, locked to the input burst. During playback, the 175fH VCO is controlled by the APC circuit (Fig. 12-12). With control only by the APC, it is possible for a mislock to occur. Such a mislock condition is prevented by an APC-ID (APC identification) circuit within Q4.

Figure 12-13 shows operation of the APC-ID circuits. The 175fH VCO output is applied through a gate to a counter. This counter is switched open and then reset every 4H by pulses from a flip-flop(FF) locked to the playback H-sync. If the 175fH frequency is correct, the counter counts 700 pulses before

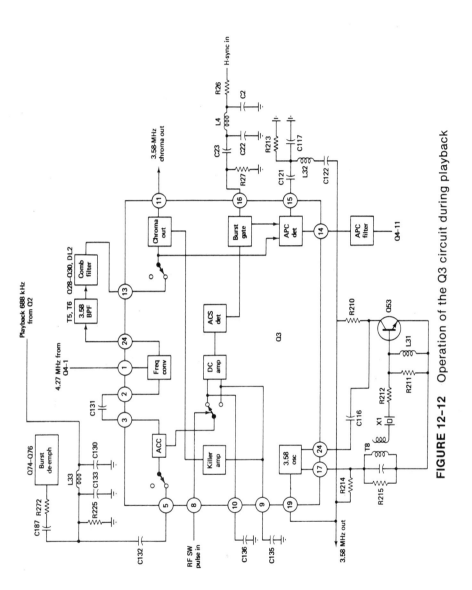

FIGURE 12-12 Operation of the Q3 circuit during playback

272   Video Circuits

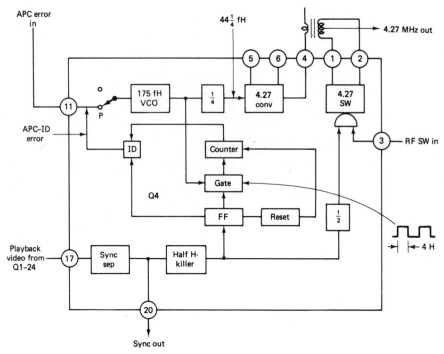

FIGURE 12-13   Operation of the APC-ID circuits

reset (175 × 4 = 700). If the 175fH VCO is off-frequency, the count is not 700 and an error voltage is applied to the input of the 175fH VCO (to correct the frequency). So the 175fH VCO is locked to both the playback 3.58-MHz (through the APC, Fig. 12-12) and the playback H-sync (through the APC-ID, Fig. 12-13).

12-3.6 Burst deemphasis circuit

Details of the burst deemphasis circuit of Fig. 12-12 are shown in Fig. 12-14. This circuit operates only during playback, when BIII is selected, and functions to attenuate the 688-kHz signal passing to pin 5 of Q3. A burst gate pulse (H-sync) is applied to the base of Q75, and a BII/BIII switching control signal is applied to the base of Q76. The switching control signal is low for BIII and high for BII. A high during BII turns on Q76, shorting the collector to emitter, and bypassing the burst gate pulse to ground. Under these conditions, Q74 and Q75 are off. This turns on Q74 only during BIII, and only during the time of color burst, so the signal is attenuated only during this time period.

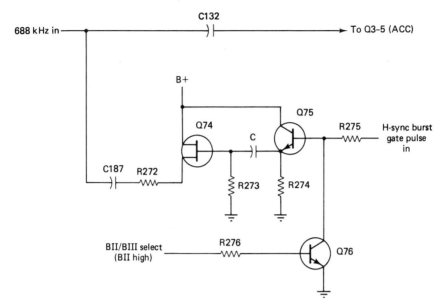

FIGURE 12-14 Operation of burst-deemphasis circuit

12-4 LUMINANCE (BLACK-AND-WHITE) TROUBLESHOOTING/REPAIR

The first step in troubleshooting luminance circuits is to play back a *known good tape* or an alignment tape. (This is not a bad idea when troubleshooting any VCR circuit!) The tape playback identifies the problem as playback or record, or both. Next run through the electrical adjustments that apply to luminance (or Y or picture), using the manufacturer's procedures. Keep the following points in mind when checking performance and making adjustments.

If playback from a known good tape has poor resolution (picture lacks sharpness), look for an improperly adjusted noise canceller and for bad response in the video head preamps. When making adjustments, study the stair-step or color-bar signals (Fig. 2-3) for any transients at the leading edges of the white bars.

If the playback has excessive snow (electrical noise), try adjustment of the tracking control. Mistracking can cause snow noise. Next try cleaning the video heads (Fig. 2-7) before making any extensive adjustments. (Cleaning the video heads clears up about 50% of all noise or snow problems.) Keep in mind that snow noise can result from mistracking caused by a mechanical problem. For example, if there is any misadjustment in the tape path, snow can result. So if you have an excessive noise problem that cannot be corrected by tracking adjustment, head cleaning, or electrical adjustment, try mechanical adjustments, starting with the tape path.

If playback of a known good tape produces smudges on the leading edge of the white parts of a test pattern (from an alignment tape) or on a picture, the problem is probably in the preamps or in an adjustment that match the heads to the preamps. The head/preamp combination is not reproducing the high end (near 5 MHz) of the video signals. The adjustment procedures usually show the head/preamp characteristics.

If you see a herringbone (beat pattern in the playback of a known good tape, look for carrier leak. There is probably some unbalance condition in the FM demodulators or limiters, allowing the original carrier to pass through the demodulation process. If very excessive carrier passes through the demodulator, you may get a negative picture (blacks are white and vice versa). Recheck all carrier leak adjustments.

Most luminance adjustment procedures include a check of the *video output level* (typically 1 V p-p). If the VCR produces the correct output level when playing back an alignment tape, but not from a tape recorded on the VCR, you may have a problem in the record circuits. As an example, the record current may be low. (One common symptom of low record current is snow or excessive noise.) Another common problem in the record circuits is the *white-clip adjustment*. Most luminance adjustment procedures include both a white-clip adjustment and a record current check.

To sum up luminance troubleshooting, if you play back an alignment tape, or at least a known good tape, and follow this with head cleaning and a check of the recommended alignment procedures, you should have no difficulty in locating most black-and-white picture problems.

12-5 CHROMA TROUBLESHOOTING/REPAIR

As in the case of luminance circuits, the color circuits of a VCR are very complex but not necessarily difficult to troubleshoot (nor do they fail as frequently as the mechanical section). Again, the first step in color-circuit troubleshooting is to play back an alignment tape, followed by a check of all adjustments pertaining to color. As in the luminance circuits, when performing the adjustment procedures, you are tracking the signal through the color circuits.

There are two points to remember when making the checks. First, most color circuits are contained within ICs, possibly the same ICs as the luminance circuits. Similarly, the color and luminance circuits are interrelated. If you find correct inputs and power to an IC, but an absent or abnormal output,

Chroma Troubleshooting/Repair 275

you must replace the IC. A possible exception in the color circuits are the various filters and traps located outside the IC on many VCRs.

Second, in most VCRs, the fixed input to the color converters comes from the same source for both record and playback (from crystal-controlled oscillators). If you get good color on playback but not on record, the problem is definitely in the record circuits. However, if you get no color on playback of a known good tape, the problem can be in either the color playback circuits or in the common fixed-signal source. So a good place to start color circuit signal tracing is to check any common source signal. For example, check any AFC circuit signals (629 or 688 kHz) and any APC signals (3.57 or 3.58 MHz). If any of these signals are missing (or abnormal), the color will be absent or abnormal.

The following notes describe some typical VCR color circuit failure symptoms, together with some possible causes.

If the hue control of the TV must be reset when playing back a tape that has just been recorded, check the color subcarrier frequency. Use a frequency counter.

If you get a "barber pole" effect, indicating a loss of color lock, the AFC circuits are probably at fault. Check that the AFC circuit is receiving the H-sync pulses and that the VCO is nearly on-frequency, even without the correction circuit.

If you get bands of color several lines wide on saturated colors (such as alternate blue and magenta bands on the magenta bar color-bar signal), check the APC circuits as well as the 3.58-MHz oscillator frequency.

If you get the herringbone (beat) pattern during color playback, try turning the color control of the TV down to produce a black-and-white picture. If the herringbone pattern is removed on black and white, but reappears when the color control is turned back up, look for leakage in both the color and luminance circuits. For example, there could be a carrier leak from the FM luminance section beating with the color signals, or there could be leakage of the 4.27-MHz signal into the output video.

If you get flickering of the color during playback, look for failure of the ACC system. It is also possible that one video head is bad (or that the preamps are not balanced), but such conditions show up as a problem in black-and-white operation.

If you have what appears to be severe color flicker on a Beta VCR, you may be losing color on every other field. This can occur if the phase of the 4.27-MHz reference signal is not shifted 180° at the H-sync rate when one head is making its pass. The opposite head works normally, making the picture appear at a 30-Hz rate.

If you lose color after a noticeable dropout, look for problems in the burst ID circuit. It is possible that the phase-reversal circuits have locked up on the wrong mode after a dropout. In that case, the color signals have the wrong phase relation from line to line, and the comb filter is cancelling all color signals. Check both inputs (3.58-MHz input from the reference oscillator and the video input signal) applied to the burst ID circuit. If the two signals are present, check that the burst ID pulses are applied to the switch-over FF.

Again, keep in mind that all the color circuit functions discussed here may be contained within one or two ICs and cannot be checked individually. So you must check inputs, outputs, and power sources to the IC, and then end up replacing the IC!

13

SPECIAL FEATURES AND CIRCUITS

The term *special features* can be applied to many circuits and functions in a modern VCR, when compared to early-model VCRs. Virtually any of the features found on modern VCRs are special, in that there are no comparable features in early-model VCRs (or the early-model features are very elementary by comparison). So, instead of trying to cover every possible feature, we concentrate on the *timers, counters,* and *displays* found in most modern VCRs. We also cover such features as *fine editing* and *index operation* found on some VCRs.

13-1 TYPICAL TIMER AND BACKUP POWER SUPPLY

As discussed in Chapter 1, the timer function permits the VCR to record program material at specific times on specific days, while not attended by the user. Most modern VCRs use a microprocessor IC as the heart of the timer function. The timer IC also assumes many additional duties over and above that of automatic program recording. Some typical functions for the timer IC include: main and remote power on-off, remote control, channel lock, one-touch recording (OTR), time of day, and control of the front-panel display.

In this section we discuss the timer circuit of a typical VCR (that shown in Fig. 1-1a). The basic timer circuit is shown in Fig. 13-1. Note that the timer microprocessor IC8A1 is a 48-pin LSI device. IC8A1 receives operational power at three inputs. The VSS inputs, pins 20 and 21, receive 9 V from a backup power supply (described in Sec. 13-1.8). The VP input, pin 23, accepts

278   Special Features and Circuits

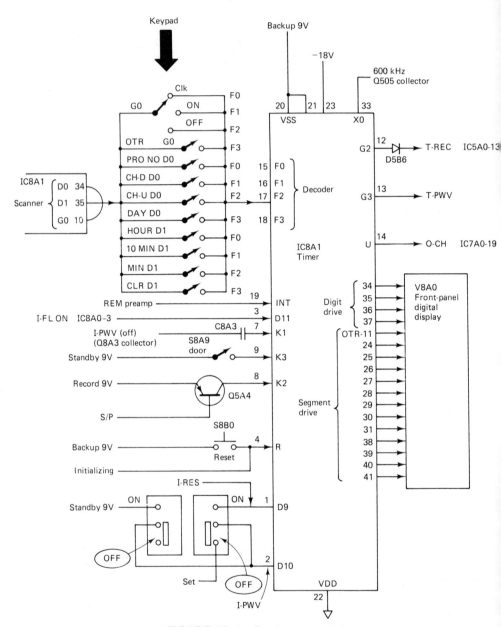

FIGURE 13-1   Basic timer circuit

−18 V from the main switched power supply. Substrate ground is provided at the VDD terminal, pin 22.

Before we get into the basic function of program recording, let us consider two requirements for any timer microprocessor, *reset* and *timing* or *time base*.

13-1.1 Timer microprocessor reset

Figure 13-2 shows the basic circuit for reset of the timer microprocessor. Reset establishes a starting point, forcing all outputs of the microprocessor low. When reset occurs, all externally programmed memories are erased. Generally, reset is required only when power is first applied, or when power has been interrupted for a certain period of time and then regained. Without some form of reset, under these conditions, the state of the microprocessor is completely unpredictable, as are the circuits controlled by the microprocessor. A timer microprocessor can also be upset by momentarily losing pulse, very short power interruption, or even a timing signal "spike." Because of the problems, IC8A1 is provided with two reset functions, *automatic* and *manual*.

When power is initially supplied to the VCR, the 9-V backup power supply is energized. As this voltage begins to rise to the full 9 V, a pulse is differentiated by C505 and applied to pin 4 of IC8A1 as a high to reset the microprocessor. This is an automatic function during initial turn-on.

If power is lost for a period of time exceeding the maintenance capabilities of the backup 9-V supply (when the internal batteries maintaining operation of the circuit begin to discharge) and the 9-V potential begins to decrease (approaching the 7-V level), Q503 turns off, turning on Q504 and applying a reset high to pin 4 of IC8A1. Again, this is an automatic function that resets all the IC8A1 outputs low and erases all external memories.

Manual reset is provided by S8BO, the front-panel reset switch. When S8BO is pressed, backup 9 V is applied directly to pin 4 of IC8A1. The manual reset can be used any time. Generally, manual reset is used when the automatic reset fails (due to a defect) or when there is a short-term problem beyond the capabilities of the automatic reset (such as power interruption for a fraction of a second, abnormal pulse spikes, etc.).

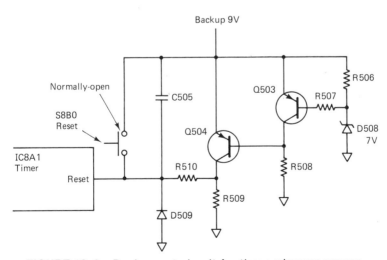

FIGURE 13-2 Basic reset circuit for timer microprocessor

13-1.2 Timing and time-base

Figure 13-3 shows the time-base generator circuit for IC8A1. Timing is necessary to synchronize the internal operations of any microprocessor. Timer microprocessors such as IC8A1 also rely on timing signals to supply the *reference* necessary for *time-of-day* operations. These timing signals are provided by IC501, which is a combination crystal-controlled oscillator and frequency divider.

IC501 generates three square-wave clock pulses (with a 50% duty cycle) at three specific frequencies: 300, 600, and 1200 kHz. Only the 600-kHz timing signals are used by IC8A1. The timing signals are highly stable and temperature-compensated by IC501 to maintain a frequency stability of ±5 ppm at room temperature.

The 5-V 600-kHz timing signal at the output of IC501 (pin 1) is amplified to about 9 V by Q505 and applied to the XO input of IC8A1 (Fig. 13-1). Dividers within IC8A1 reduce the 600-kHz frequency to an appropriate value for the time-of-day reference.

13-1.3 Programming record function

The program timer and record functions shown in Fig. 13-1 are essentially the same for all VCRs. That is, specific scanner signals are applied to appropriate decoder inputs. These inputs instruct the microprocessor to accept data relating to a particular part of the programming process. The VCR of Fig. 13-1 has a capacity of eight complete programs, each program consisting of: (1) *start time,* (2) *end time,* (3) *day of week* (within a 14-day period, or every

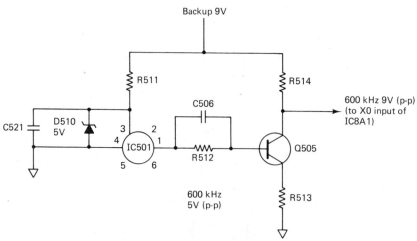

FIGURE 13-3 Time-base generator circuit

day), and (4) *channel data*. IC8A1 accepts data relative to any of these program segments entered in any sequence from the *key pad*.

The key pad forms a portion of the *key matrix* input and can be subdivided into two sections, as shown in Fig. 13-1.

(1) *Mode switch: Clock on* (start time) and *off* (end time), are used to notify IC8A1 to select the appropriate memory cell that is to accept information. Each switch position must be maintained during a phase of programming. Pressing a mode switch also displays the contents of the memory section related to that particular mode-switch position.

(2) *Increment keys:* (Day, hour, 10 minute, minute, channel up, channel down, and program number) are used to write incremented data into the IC8A1 memory cell selected by the mode switch.

The key pad input lines accept microprocessor-originated scanner signals (a form of oscillator-derived signal, each at a specific phase) and form IC8A1 terminals 35(D1), 34(DO), and 10(GO). Each scanner signal is applied to a particular key pad switch configuration. For example, the DO scanner signal is applied to the *program number* (S8B4), *channel-down* (S8B5), *channel-up* (S8B6), and *day* (S8B7) switches simultaneously. However, the opposing contact of each switch is connected to a separate IC8A1 decoder input. For example, program number is connected to the FO input, pin 15, while day is connected to the F3 input, pin 18. So closure of any switch on the key pad connects a scanner signal to a specific decoder, producing a "signature" or combination recognized by IC8A1.

Example of a typical record programming operation. The following steps outline the sequence for a typical record programming operation.

1. A program location is selected by setting the mode switch to on (start time) position. A program number (1-8) appears at the lower right of the front-panel digital display V8A0 (Sec. 13-3). If the program number displayed contains programming information, the start time also appears on the display. If not, and the program is available, the letters "EE: EE" appear. By pressing the program-number key, the DO scanner signal is applied to the FO decoder input, and IC8A1 increments the program numbers, one at a time, from 1 to 8. The results of the incrementation appear on the display. When the desired program number is reached, the program-number key is released.

2. The day of week is then selected by pressing and holding the day key, connecting a DO scanner signal to the F3 decoder input. The day-of-week advances, one day at a time, with the results indicated on the display. When Sunday is reached and passed, the WK-2 indicator turns on, indicating the second week. When Sunday is again reached and passed, all day-of-week indicators turn on simultaneously, denoting that everyday programming is selected. Releasing the

day key when the desired day of week is indicated discontinues the incrementation and enters the displayed day of week into the memory cell of the selected program location.

3. Start time is then selected by pressing the *hour* (D1 to F0), *10 min* (D1 to F1), and *minute* (D1 to F2) keys.
4. The channel is selected by pressing either the channel-up (D0 to F2) or channel-down (D0 to F1) keys. The channel increments up or down, and the results are displayed on the LED channel display. Releasing either button enters the displayed channel into memory.
5. The end time is then selected by placing the mode switch in off (G0 to F2). Step 3 is then repeated to place the end time into memory.

If an error is made during the programming process, only the erroneous instructions need be rewritten. If you want to erase the entire contents of a particular program location without affecting the remaining program locations, select the appropriate program location and press the clear key (D1 to F3). Only the program location number indicated on the display is then affected.

Programming is not accepted by IC8A1 unless a high exists at the K3 input, pin 9. The door switch S8A9 controls this action by applying standby 9 V to pin 9 through the closed contacts of S8A9 when the *program door* is open. Since standby 9 V is used, programming may commence even though the VCR is turned off.

IC8A1 ignores programming attempts under certain conditions. For example, when the *reserve* switch S8A7 is on, 9-V power is applied at the D9 input, pin 1. Also, IC8A1 ignores programming instructions during the execution of a timed-record program or during an active record mode.

13-1.4 Channel lock during record

To prevent the inadvertent interruption of the recording process, channel selection is inhibited during the active record mode. This is called *channel lock*. The remote power supply control and programming are also inhibited by channel lock. During any record mode, record 9 V is applied to the emitter Q5A4 as shown in Fig. 13-1. The base of Q5A4 is controlled by still/pause (S/P) logic (which is high in S/P). So Q5A4 conducts during record, placing 9-V power at the K2 input, pin 8. Should the S/P mode be engaged, the high on the base of Q5A4 turns off Q5A4, removing the 9-V power from pin 8 and cancelling the inhibit instruction.

13-1.5 Program execution sequence

The execution of a program record operation is essentially a three-step procedure:

1. IC8A1 outputs a high on the T-PWV line G3, pin 13, enabling the switched-B+ supplies and placing the VCR in an active-on condition.
2. A channel-select signal is output on the O-CH line U, pin 14, and is directed to the FS tuning system microprocessor IC7A0 (Chapter 10), selecting a preprogrammed channel.
3. A high is generated on the T-REC line G2, pin 12, routed to the input expander IC5A10, placing the VCR in the record mode.

The three actions do not occur simultaneously, but in a timed sequence. A certain amount of time is required to complete the channel-selection process, allow the tape to be withdrawn from the cassette and loaded into the transport, and to compensate for the small amount of recording time lost due to the delay inherent in the automatic fine-editing process executed at the beginning of any active record mode (as discussed in Sec. 13-6). To compensate for these cumulative time losses, the internal programming of IC8A1 is such that the program-record operation begins approximately 6 seconds prior to the preprogrammed start time.

The channel-selection procedure begins immediately. About 1.2 seconds later, the switched power supplies are energized. An additional 2 seconds later, the record mode is selected. Following tape loading and fine editing, recording begins at the preselected start time.

13-1.6 One-touch recording (OTR)

OTR, also a function of IC8A1, is basically a manually operated record operation, producing a sequence of events essentially identical to programmed record operation. OTR permits the user to initiate the record mode and record from 30 minutes to two hours. Pressing the OTR button once produces 30 minutes of recording. Pressing OTR twice yields an hour, three times produces 90 minutes, and four times produces two full hours of recording. If OTR is pressed a fifth time, in sequence, recording time reverts back to the original zero time.

The selected OTR time is indicated on the display (Sec. 13-3). When the recording process begins, the time is decremented in one-minute steps, displaying the OTR time remaining. Within five seconds after the display indicates "0-00" OTR time remaining, the VCR is turned off automatically.

The six-second delay prior to recording (described in Sec. 13-1.5) is also true for OTR. The sequence is as follows:

1. A high is output on the O-OTR line G1, pin 11, and turns on the OTR indicator of display V8A0. Both the segment and digit drive from IC8A1 is then the OTR time selected and time remaining (rather than any programmed time indications).

2. Approximately 3.2 seconds later, a high is output at the T-REC line G2, pin 12, placing the VCR in the record mode (as in step 3 of normal program execution, Sec. 13-1.5).

13-1.7 Power on-off during record timing

Timer IC8A1 is an integral part of the power on-off operation. A power-on condition is produced by energizing various switched-B+ supplies (Chapter 3). The B+ switching circuits are controlled by logic from IC8A1 developed on the T-PWV line G3, pin 13. (This line is used to enable the switched-B+ supplies at the beginning of a program record sequence, and is an internal function of IC8A1.)

Logic on the T-PWV line can also be determined by power switch S8A8 and reserve switch S8A7. When both switches are off, the I-PWV and I-RES lines are low. All power-on functions are inhibited, including the program-record power-on and remote-control power-on functions. Under these conditions, the VCR is not operational, and only the time-of-day display remains active.

Placing power switch S8A8 on directs standby 9-V power to the I-PWV line. IC8A1 responds by generating a high on the T-PWV line, enabling the switched-B+ supplies. In this operational state, the remote-control power on-off control can be used to control logic on the T-PWV line (via the remote-control input at the INT terminal, pin 19, as described in Chapter 4).

Pressing the remote-control power on-off key alternately toggles the T-PWV output from a high to a low, and vice versa. The program-record on-off remains inhibited.

When power switch S8A8 is on, placing the reserve switch S8A7 on applies standby 9-V power to the I-RES line. The T-PWV line immediately drops to a low, and the switched-B+ supplies are disabled. Under these conditions, the remote-control power on-off input is inhibited, and the VCR is in (essentially) the off state. (That is, the VCR is under complete control of IC8A1, and a power-on instruction of the T-PWV line, if any, depends solely on the program-record start times.)

Keep the following points in mind when troubleshooting a timer system such as shown in Fig. 13-1:

- The power switch must be on to enable the remote-control power on-off capability.
- Power and reserve switches must be on before programmed recording can proceed.
- When the reserve switch is on, timer IC8A1 does not recognize commands issued by either the remote control or the power on-off switch.

13-1.8 Backup power supply for timer

If ac power to the VCR is lost completely as the result of a temporary power failure, the backup power supply circuit of Fig. 13-4 takes over. The backup supply is powered by a 2.6-V rechargeable battery, and develops a backup 9 V to the timer and programming circuits for periods up to 30 minutes following power interruption. This maintains the clock and programmed memories. When power is regained, the memories are intact and the battery recharges. Approximately 24 hours is required to recharge the battery from a fully discharged state. Even though memories are retained, should a power loss occur during an operational mode (record, playback, etc.), the VCR *unloads automatically* when power is regained.

When ac power is interrupted, the 2.6-V battery supplies bias voltage to Q501 (which is connected to T501 to form a blocking oscillator). Oscillations from Q501 are rectified by D505/D507, regulated by D506, and filtered by C503/C504/L501. The resulting dc voltage is available to all circuits powered by the backup 9-V voltage source. Approximately 30 minutes is required for the battery charge to deplete.

When ac power is recovered, the nonregulated 15-17 V turns Q502 on through D503. The conduction of Q502 grounds the base of Q501, preventing oscillation. The nonregulated power is also directed through current-limiting resistor R501 and diode D502, providing a 20-mA charging current for the battery.

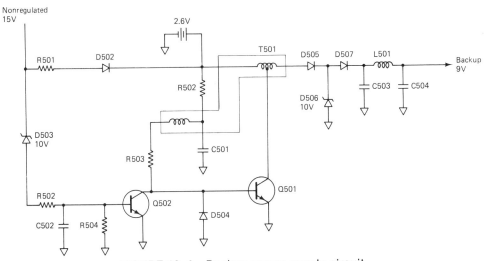

FIGURE 13-4 Backup power supply circuit

13-2 AN ALTERNATE TIMER-IC/KEYBOARD/DISPLAY CONFIGURATION

Figure 13-5 shows the timer-IC/keyboard/display configuration for a typical VCR (that shown in Fig. 1-1c). The timer microprocessor IC101 monitors all input/output commands of the tuner/timer/charger, drives the multidigit front-panel display, performs timekeeping functions and timer record operations. Compare this to the timer microprocessor shown in Fig. 13-1. There are many similarities, but the two microprocessors are definitely not the same. For example, IC101 receives B+ power from a 5-V *backup supply line*. This line has a *0.22-farad* (not microfarad) capacitor as a filter. The capacitor maintains power to IC101 in the event of a power failure. The discharge rate of the capacitor is over one hour.

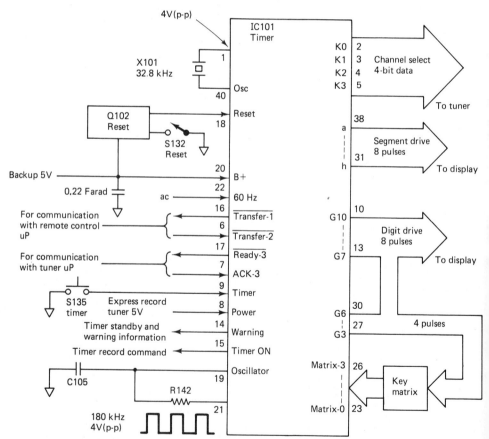

FIGURE 13-5 Typical timer-IC/keyboard/display configuration

An Alternate Timer-IC/Keyboard/Display Configuration 287

During normal operation, the 60-Hz ac signal applied to pin 22 of IC101 is used to maintain proper time. In the event of a power failure (for the memory backup period), an internal 32.8-kHz oscillator at pins 1 and 40 are used for timekeeping. The long-term accuracy of the 60-Hz signal is far better than the 32.8-kHz signal. However, during the one-hour battery backup period, the 32.8-kHz signal is sufficiently accurate to maintain timing.

IC101 generates eight scanning pulses, G3 through G10, to drive the display and key matrix circuit. In this case, the key matrix circuits consist of a series of switches connected to generate four possible outputs (matrix 0 through matrix 3, at pins 23 through 26).

The key matrix includes an *express record switch* which turns on an express record circuit within IC101. The express record circuit allows *immediate recording* for 30, 60, or 90 minutes, or for 2-, 3-, or 4-hour periods. The express-record enable signal is applied to pin 8 of IC101 (which is the divided-down tuner 12-V signal). During stop mode, express record can be turned on. In play, the enable signal is not present at pin 8, and express record is inhibited. During time-recording operation, pins 9, 14, and 15 are used to control the VCR, as discussed in Sec. 13-2.1.

Timer IC101 selects the desired channel during timer-record operations by applying a 4-bit channel-select signal at pins 2 through 5, the K0 through K3 outputs (as discussed in Sec. 10-3 of Chapter 10). Two groups of information are passed from IC101. The first group is the tens digit of the channel required, while the second group is the ones digit. The two groups are passed to the tuning PLL microprocessor (Fig. 10-15) to select the proper channel during the timed recording mode.

In order for IC101 to communicate with the tuning PLL microprocessor, various handshaking lines (at pins 6, 7, 16, and 17) are used to gain control of the 4-bit data bus (as discussed in Sec. 10-3.1 and shown in Fig. 10-16).

Timer IC/keyboard/display troubleshooting/repair. If you do not get timed recording (the function is absent or abnormal) or there is an improper display during timed recording, try correcting the condition by pressing the *reset* switch S132. This should reset the timer microprocessor as well as the display. If not, check the timer and related circuits as described in Sec. 13-2.1. Before you get into timer circuit troubleshooting, first check for correct power to IC101 at pin 20, a proper reset signal at pin 18 (when reset is pressed), and for 32.8-kHz oscillator signals at pin 1 (which should be about 4 V p-p for IC101).

If all these signals are present, but either the timer or display functions are not normal, then go into the circuits as described in Sec. 13-2.1. Keep in mind that the timer, display, tape counter, and system-control circuits of most modern VCRs are interrelated, and this should be considered when troubleshooting any of these circuits. Troubleshooting for display/counter circuits is discussed further in Secs. 13-3 and 13-4, as well as Chapter 14.

13-2.1 Timed record operation

Timed record operation is controlled by timer IC101 as shown in Fig. 13-6. After the proper record time is programmed into IC101 and timer switch S135 is pressed, a low is applied to pin 9. At this time, one of three signals appears at pin 14. Dependent upon the previous state of the signal at pin 14, the signal can be *steady low, steady high,* or *warning* (1 Hz).

If the timer operation is off, and the timer switch S135 is pressed, the signal at pin 14 goes steady high (standby condition), or a 1-Hz square wave (warning condition) is generated. The 1-Hz square wave warning is present when there is a cassette tape within the VCR that has the record tab broken out, or when no cassette is present, as discussed in Sec. 1-9.

The signal at pin 14 drives Q101, which turns on the timer LED. If the

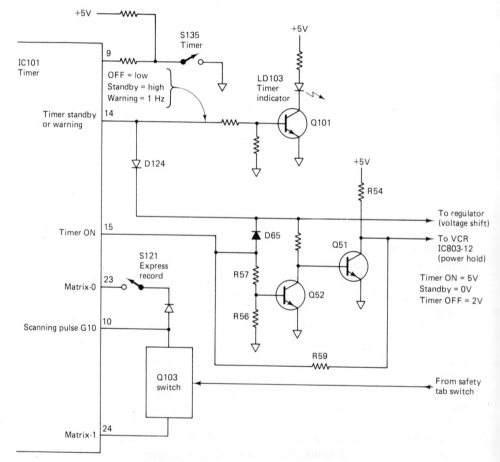

FIGURE 13-6 Timed-record operation

An Alternate Timer-IC/Keyboard/Display Configuration 289

warning signal is present, then Q101 is turned on and off at the 1-Hz rate, causing the timer LED to blink on and off. This alerts the user of a problem with the cassette (either no cassette or a cassette that is not to be recorded upon).

With a cassette installed (with record tab in place), the signal at pin 14 is high. This high is passed through D124 to bias the VCR control circuits, consisting of Q51 and Q52. The standby high is applied to the collector of Q52 and the base of Q51.

During the standby mode, (after S135 is pressed, but before IC101 starts the record mode) the signal at pin 15 of IC101 is low. This low is applied to the base of Q52, keeping Q52 off and Q51 on. With Q51 on, the collector of Q51 goes low. This low is applied to the power-hold IC803 (which is part of system control as described in Chapter 14).

When IC101 determines that the recording should start (after S135 is pressed and after the standby period), pin 15 goes high, turning Q52 on and Q51 off. With Q51 off, the collector of Q51 goes high. This high signals system control to go into record.

During the period when S135 is not pressed (the off condition, before standby or record), pins 14 and 15 are low. With pin 14 low, both Q51 and Q52 are off. This develops approximately 2 V at the collector of Q51, and signals system control that timer IC101 is off. (The 2 V is produced by the voltage divider R54/R59, which is returned to ground through pin 15 of IC101.)

The express-record switch S121 applies a G10 scanning pulse to matrix-0 at pin 23. The G10 scanning pulse is also applied to matrix-1 input, pin 24, during the time when the safety tab switch is activated. (This occurs when the safety tab has been removed from the cassette or when the cassette is not in place, Sec. 1-9.)

Timed record troubleshooting/repair. As discussed, the first step in troubleshooting any timer or timer/display function is to press the reset switch S132. (This may cure the trouble!) Next check that you have properly entered all timed recording and channel information at the key matrix. Then press the timer switch S135 and check for a low at pin 9 of IC101. If the low is missing, check S135 and the related wiring.

Next check the signal at pin 14 of IC101. If the signal is a 1-Hz square wave, suspect a problem in the cassette safety switch area or transistor Q103. If the signal at pin 14 is low, press the timer switch S135 once and see if pin 14 goes high. If the logic condition of pin 14 does not change when S135 is pressed, suspect IC101. If pin 14 goes high when S135 is pressed, check that the collector of Q51 is low or zero volts. If not, suspect Q51/Q52.

Next check the signal at pin 15 of IC101. Pin 15 should remain low immediately after S135 is pressed, but should go high after a standby period. If pin 15 does not go high at any time after S135 is pressed, suspect IC101.

If pin 15 goes high after the standby, check that the collector of Q51 goes high (to about 5 V). If not, suspect Q51/Q52. If the collector of Q51 goes high, but timed recording does not occur, suspect the system-control circuits (Chapter 14).

13-3 TYPICAL ELECTRONIC COUNTER AND DISPLAY

Figure 13-7 shows the electronic counter and display circuit for a typical VCR (that shown in Fig. 1-1a). Note that the counter microprocessor IC8A0 uses the same front-panel digital display V8A0 as does the timer microprocessor IC8A1 (discussed in Sec. 13-1 and shown in Fig. 13-1). This is typical for most modern VCRs. Also typical is the fact that the same front-panel display is often used by system-control circuits. In any VCR, there is an interrelationship among the timer, display, counter, and system control. Keep this in mind when troubleshooting any of these circuits.

13-3.1 Electronic counter

Counter microprocessor IC8A0 includes a conventional *tape counter with memory* which drives a four-digit display to indicate the amount of tape consumption in any operational mode. The tape counter can count up or down. When the memory feature is used, a particular portion of tape can be *indexed* by resetting the counter to the 0000 position, allowing play or record to continue, and then pressing the front-panel memory switch. The rewind mode is automatically engaged, and the counter decrements. When the 0000 display indication is reached, the rewind mode is terminated, and the tape comes to a halt at the preindexed counter setting. (Another form of indexing is discussed in Sec. 13-7.)

IC8A0 calculates and displays the playback or record *time remaining* on a particular tape in the play, record, FF (fast forward), and rewind modes. IC8A0 includes a *tape-end* indicator function which turns on the front-panel display *end* indicator when time remaining is less than five minutes (at any tape speed).

Since a single fluorescent display is used for all functions requiring a display indication, IC8A0 selects the output applied to the display in accordance with the counter, or timer, function in use. IC8A0 is similar to timer IC8A1, but with a different program.

13-3.2 Tape counter with memory

The tape-counter function indicates (on a four-digit display) index numbers showing the amount of tape consumed in a given operational mode. The numbers on the display increment and decrement at the rate of one count, or digit change, for each complete revolution of the take-up (TU) reel (Chapter 5).

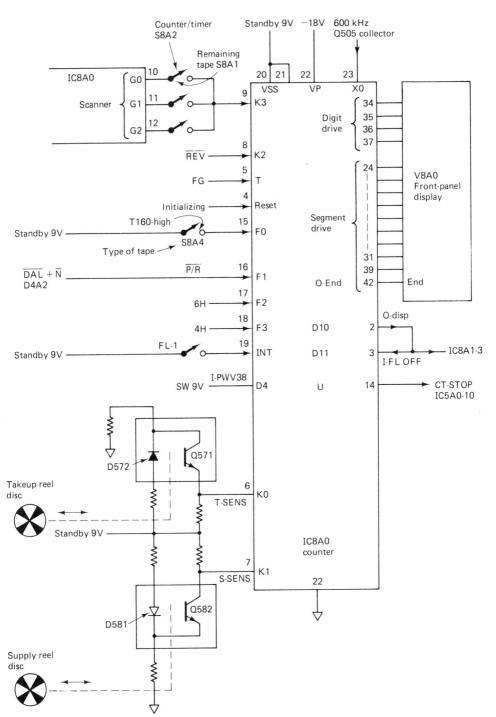

FIGURE 13-7 Typical electronic counter and display circuits

In the counter mode, IC8A0 accepts pulses generated by the T-SENS detector, which includes phototransistor Q571, LED D572, and the TU reel disc. The disc is coated with an alternating black-and-white pattern, and is driven by the TU reel. The light from LED D572 is applied to the disc surface. The white segments of the disc reflect light, while the black segments absorb the light. This produces pulses of light as the disc rotates. The pulses are picked up by phototransistor Q571 and applied to the T-SENS input K0 at pin 6 of IC8A0, at a rate of three pulses for each complete revolution of the TU reel.

The $\overline{REV}$ logic, which determines whether the counter increments (forward) or decrements (reverse), is applied to the K2 input at pin 8. $\overline{REV}$ goes high when the TU reel is rotating in a clockwise (forward) direction, and vice versa.

When counter reset switch S8A2 is pressed, the G2 output at pin 12 of IC8A0 is applied to the K3 input at pin 9. This resets the counter and drives the display to 0000. The T-SENS input at pin 6 is inhibited during counter reset.

Selection of the memory function (through action of the mechanical control, Chapter 5) automatically engages the rewind mode. When the counter reaches the 0000 display indication, IC8A0 generates a high at the U input, pin 14. The high is applied, via the CT-STOP line, to pin 10 of input expander IC5A0, terminating rewind and engaging the stop mode.

13-3.3 Remaining tape display

The remaining tape-display circuits use the following information: tape length (T-30, T-60, T-120, etc.), tape speed (2H, 4H, or 6H), operational mode (record; playback, including search and slow modes; FF; and rewind), and the volume of tape remaining on both the supply and take-up reels. This information is interpreted by IC8A0, which responds by calculating the amount of operational time remaining on a given tape and then displaying the results (in hours and minutes) on display V8A0.

The display capability, when operating in the time-remaining mode, extends for 0:00 to 9:59, with a minimum remaining time that can be accurately displayed of 0:01 (one minute).

When front-panel remaining tape switch S8A1 is pressed, the G1 output from pin 11 of IC8A0 is applied to the K3 input at pin 9. IC8A0 then accepts data at various inputs, interprets the data, and computes the remaining time, prior to generating the required outputs to display V8A0. A certain amount of time is required for interpretation and computation. The results are generally not available to the display instantly. In some cases, a full minute is required for display. Switch S8A1 must remain closed until computation has been completed and the remaining time is displayed. Display V8A0 signifies that a computation is in progress by displaying a "-:-" or WAIT symbol.

Tape speed logic, from the automatic speed selector, is applied to pins

17 and 18 of IC8A0. 6H logic (which is high only in 6H) is applied to the F2 input at pin 17, and 4H logic (which is high only in 4H) is applied to the F3 input at pin 18. In 2H, both inputs at pins 17 and 18 are low.

The logic at the F1 input, pin 16, tells IC8A0 the operational mode in progress (playback, record, FF, or rewind). The F1 input uses a $\overline{\text{PB-REC}}$ voltage developed from the $\overline{\text{DAL}}$ (low after loading) and $\overline{\text{N}}$ (low in PB and REC) voltages. The $\overline{\text{PB-REC}}$ logic is low in record, playback, or any special-effects mode that is valid in a tape-loaded condition. The $\overline{\text{PB-REC}}$ logic goes high only when the tape is unloaded (during FF or rewind).

During playback or record (low at pin 16), IC8A0 reads data at three inputs: the T input (pin 5), the K0 input (pin 6), and K1 input (pin 7).

The T input accepts capstan FG pulses (Chapter 8) taken from pin 20 of IC4A5 and amplified/inverted by Q4D7. Since the frequency of the FG signal varies in direct proportion to capstan motor speed, the FG frequency also indicates the *rate of tape consumption*.

The K0 and K1 inputs accept pulses generated (coincident with rotation of the take-up and supply reels) by the T-SENS and S-SENS detectors. (Basic operation of the S-SENS detector is the same as for the T-SENS detector, as described in Sec. 13–3.2, except that the S-SENS detector reads light pulses from the supply reel.)

Both the T-SENS and S-SENS detectors develop three pulses on their respective output lines for each complete rotation of the corresponding tape reel. However, the *duration* of each pulse depends on the *volume of tape* contained on each reel. This volume exerts an influence on the *rotational speed* of each reel. While the change in speed is very subtle, the change in pulse period or duration is sufficient for IC8A0 to perform calculations. As an example, toward the beginning of a tape, when most of the tape is on the supply reel, the take-up pulse periods are shorter than the supply pulse periods. The opposite is true toward the end of a tape. At the center of a tape run, both reels contain the same amount of tape, and the pulse periods are virtually identical. IC8A0 interprets information to all three inputs (T, K0, and K1) and produces the necessary display signals to show remaining tape time.

During FF or rewind modes (a high at pin 16), IC8A0 alters the method of calculation and responds solely to the pulse periods on the S-SENS input at pin 7 for calculation of remaining tape time.

IC8A0 automatically detects the *length of tape* in use (relative to information received on the T-SENS and S-SENS input) and alters the calculation method to coincide with the use of T-30, T-60, or T-120 tape. However, IC8A0 cannot automatically detect the use of a T-160 tape. So *tape type* switch S8A4 must be set to the T-160 position when T-160 tape is used. This places a high on pin 15 of IC8A0. Pin 15 is low (S8A4 open) for all other tape lengths.

Once the initial calculation is performed, IC8A0 continues to refresh, or to maintain, the calculation contents for the duration of the selected mode. The results of the calculations are available to display V8A0 at any time by pressing the remaining tape switch. The calculation contents are automatically cleared if the type of tape or tape speed is changed. Also, when stop or eject are selected, the contents of remaining tape calculations are cleared by a high at the INT input, pin 19, from the FL-1 switch (which is closed during loading or unloading).

Within five minutes of the end of a tape, IC8A0 outputs a series of 0.64-second pulses on the O-END line, pin 42. These pulses turn on the END indicator of display V8A0.

IC8A0 is timed by a 600-kHz signal applied at pin 33. The digit- and segment-drive signals from IC8A0 to display V8A0 appear on pins 34–37 and 24–31, respectively.

13-3.4 Counter/timer display switching

Figure 13-8 shows the counter/timer display-switching circuits. These circuits are necessary since both the counter IC8A0 and the timer IC8A1 use the same fluorescent display V8A0. IC8A0 outputs digit/segment data corresponding to the counter mode (which uses a four-digit display), as well as digit/segment logic corresponding to remaining time (a three-digit display). IC8A1 outputs digit/segment data corresponding to both the time of day and the remaining time of OTR (one-touch recording) operation.

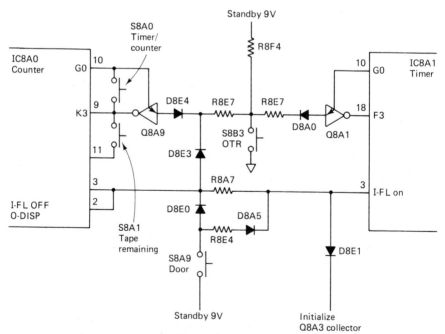

FIGURE 13-8 Counter/timer display switching circuits

Typical Electronic Counter and Display   295

The factor controlling the source of the digit/segment logic applied to display V8A0 appears on the O-DISP output, pin 2 of IC8A0. This O-DISP output is determined primarily by counter/timer switch S8A0. Closing S8A0 momentarily connects a GO output from pin 10 of IC8A0 to the K3 input at pin 9. This changes the O-DISP output from a low to a high, or vice versa. (The state of the O-DISP output changes each time S8A0 is pressed.)

In turn, the O-DISP output determines the logic at the I-FL OFF input, pin 3 of IC8A0, and at the I-FL ON input, pin 3 of IC8A1. A high at the I-FL OFF input disables the display outputs of counter IC8A0, while the same high at the I-FL ON input enables the display outputs of timer IC8A1. Under these conditions, display V8A0 indicates the time of day. If the OTR function is engaged under these conditions (timer on, counter off) by closing OTR switch S8B3, D8A0 conducts, enabling Q8A1. Conduction in Q8A1 generates digit/segment logic corresponding to the recording time remaining in the OTR mode.

No matter what logic appears on the I-FL ON input, if the program door is opened, door switch S8A9 closes, forward-biasing D8A5 and producing a high at the I-FL ON input. This enables the IC8A1 display outputs. Closure of door switch S8A9 also turns on D8E0, placing a high on the I-FL OFF input of IC8A0. This disables the counter IC8A0 display outputs (if they are in use at the time).

When counter mode is selected by S8A0, the O-DISP output goes low, turning on the counter display outputs and disabling the timer display outputs. Display V8A0 then receives digit/segment logic corresponding to the counter mode or, if the remaining time switch S8A1 is engaged, digit/segment logic corresponding to the remaining time mode.

If the OTR function is selected during the counter mode, closure of S8A3 turns on D8E4 and Q8A9. The conduction of Q8A9 simulates the closure of S8A0, driving the O-DISP output to a high and selecting the display outputs of timer IC8A1.

13-3.5 Fluorescent display

Figures 13-9 and 13-10 show the display drive and indicator drive, respectively, for the fluorescent display (of the VCR shown in Fig. 1-1a). As shown, the display circuits interface with the counter IC8A0, timer IC8A1, driver IC8A2, mechanical control IC5A1 (Chapter 5), and various discrete component networks. The counter/timer portion of the display is a dynamic type, accepting digit drive/segment logic from IC8A0 and IC8A1. The mode portion of the display is a static type where each mode indicator is turned on by a high (during the corresponding mode) applied to the associated display input.

Driver IC8A2 accepts logic from the automatic speed selector, Dolby switch, FL-1 (forward-loading switch 1), and the humidity sensor to indicate tape speed, Dolby N/R, cassette-in, and dew or humidity condition, respec-

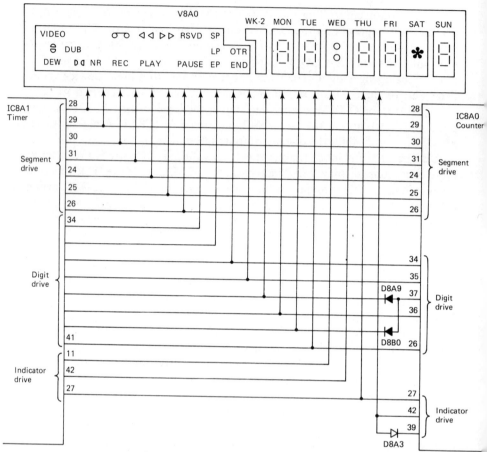

FIGURE 13-9 Fluorescent display drive

tively. Individual tape mode indicators, including TV/video and audio dub, are turned on by logic originating from the S and D output ports of mechanical control IC5A1.

The RSVD indicator is turned on by logic from Q8A0, in conjunction with the power and reserve switches. With both switches on, switched 9-V power is no longer available at the base of Q8A0, keeping Q8A0 on and generating a high at pin 14, the RSVD input.

Filament-bias voltage for the fluorescent tubes, applied to pins 1, 2, 56, and 57 of V8A0, is taken from the −18-V supply through dimmer switch S8C2, as shown in Fig. 13-11. The position of S8C2, in association with zener diodes D8D2 and D8D3, varies the applied filament bias voltage, reducing or increasing the intensity of the display.

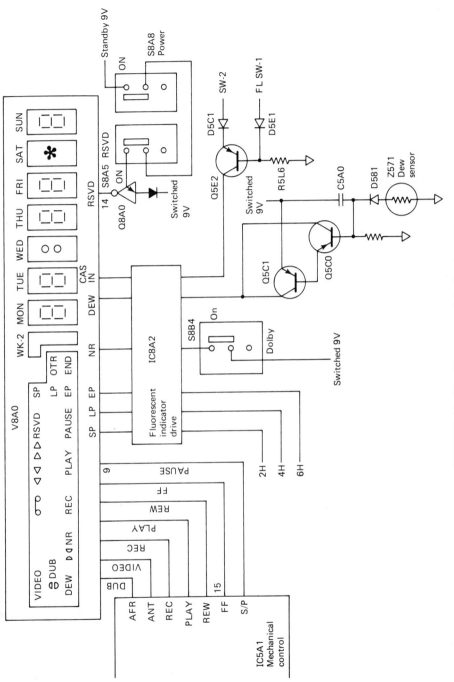

FIGURE 13-10 Fluorescent indicator drive

298 Special Features and Circuits

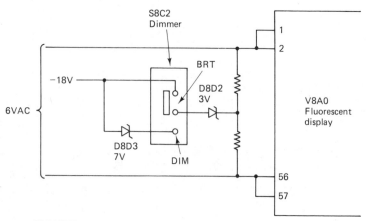

FIGURE 13-11 Fluorescent display-dimmer circuits

Display troubleshooting/repair. Troubleshooting VCR display circuits is usually not difficult. Also, many VCR displays are not serviceable. The trend is toward a single display tube or assembly (generally fluorescent or LCD), which you replace as a unit.

Before you replace any display, always check that the display is receiving power. If any of the display is on, the power is probably good. If only a portion of the display is bad, check that the corresponding signal is available. For example, if you do not get a fast-forward (▶▶) display, but other indications are good, check for a signal from pin 15 of the mechanical control IC5A1 to pin 9 of display V8A0, as shown in Fig. 13-10. Of course, if the function is not available (no fast-forward operation, in this case) do not expect the display to be on. However, there are cases where the display is good, but the function is absent or abnormal.

13-4 AN ALTERNATE TAPE COUNTER/TIME REMAINING CONFIGURATION

Figure 13-12 shows the tape-counter and time-remaining circuits for the VCR shown in Fig. 1-1f. In this case, a single microprocessor IC171 is used for both the counter and time-remaining functions. Note that the configuration described in this section performs the same overall functions as previously described, but in quite a different manner. This alternate configuration is included here to show you one of the many variations found in modern VCR circuits, and *the necessity to study the service literature for the VCR you are servicing.* Since the VCR described here is an RCA (the VJT700), we use RCA cassette sizes for reference.

An Alternate Tape Counter/Time Remaining Configuration 299

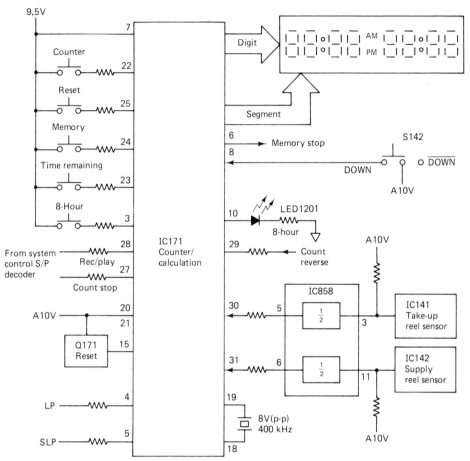

FIGURE 13-12 Tape-counter and time-remaining circuits

13-4.1 The calculator and sensors

Four sizes of RCA cassette tapes are available: VK065, VK125, VK250, and VK330. There are two characteristics that distinguish the difference between these four types: *thickness* of the tape and *diameter* of the supply and take-up reels. The VK065 and VK125 cassettes have normal-thickness tape and larger-diameter reels. The VK250 has normal-thickness tape, but smaller-diameter reels. The VK330 has the same smaller-diameter reels as the VK250, but even thinner tape. IC171 monitors the supply and take-up reel rotation speed and calculates the type of cassette in the VCR. IC171 then displays the time remaining on a front-panel four-digit tape counter.

Both the supply and take-up reels contain a *Hall-effect* detector to monitor the rotation speed of both reels, as shown in Fig. 13-13. By supplying

300 Special Features and Circuits

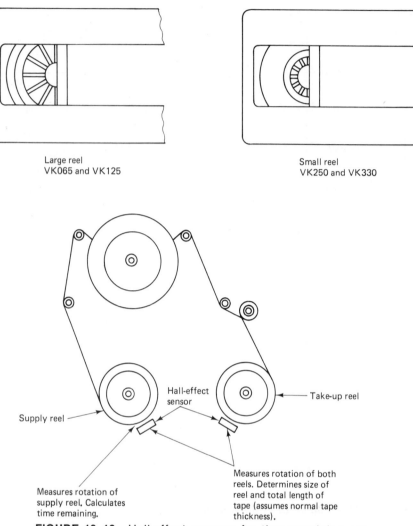

FIGURE 13-13 Hall-effect sensors for time-remaining calculation and tape-counter function

Hall-effect signals to IC171, the time remaining on the cassette tape can be calculated. IC171 has two modes of operation: calculation and display.

During the calculation mode, the four-digit display generates a series of four hyphens which blink on and off in sequence to indicate the direction the tape is moving. For example, in play, the hyphens blink on in sequence from left to right, indicating that the tape is moving from the supply reel to the take-up reel. In reverse, the hypens blink on from right to left, indicating reverse tape movement. During calculation, IC171 automatically determines the size of the reels and total length of the tape for VK065, VK125, and VK250 cassettes. For VK330 tape, the user must press an eight-hour switch to inform IC171 that VK330 tape is being used. After the correct reel size and tape length are determined, IC171 then monitors *the proportion of rotation speed* between the supply and take-up reels and calculates the time remaining on the supply reel (based on the Hall-effect signals).

During the display mode, after the time remaining has been calculated, IC171 shows the time on the four-digit display. The time-remaining display is updated continuously during playback, record, rewind, fast-forward, and search modes.

Note that the VCR must be in play or record to initialize the calculation mode of operation. If rewind or fast forward are selected when the cassette is first inserted, the calculator mode does not operate, but simply displays tape direction movement (sequencing hyphens).

13-4.2 Tape counter operation

When the VCR is turned on, 9.5-V power is applied to pin 7 of IC171, enabling the output drivers to power the four-digit, seven-segment display. Also, when initial power is applied to IC171, at pins 20 and 21, and to Q171, a reset pulse is generated automatically and applied to pin 15, as shown in Fig. 13-12. This pulse resets all internal counters and initializes IC171 to start operation in the time-remaining mode (Sec. 13-4.3). IC171 can also be reset manually by the front-panel reset switch connected to pin 25.

Either tape-counter mode or time-remaining mode can be selected by the corresponding front-panel switch (counter, pin 22, or time remaining, pin 23). IC171 returns to the mode that the VCR was in (counter or time-remaining) when power was turned off.

In the play mode, IC141 generates pulses in proportion to the revolutions of the take-up reel. These pulses are divided by 2 within IC858 and applied to pin 30 of IC171. In normal (forward) playback, the count-reverse input at pin 29 of IC171 goes low. This tells IC171 to count the reel pulses at pin 30 in the upward direction (incrementing). In reverse play, rewind, or any reverse tape mode, the count-reverse line at pin 29 is high, instructing IC171 to decrement the count.

Note that in the tape counter mode, the supply-reel sensor signal at pin 31 of IC171 (from IC142 through IC858) is not used.

Tape counter troubleshooting/repair. If a malfunction exists during the tape counter mode of IC171, but the time-remaining function is normal, first check that pin 22 of IC171 goes high when the front-panel counter switch is pressed. If not, check the switch and related wiring. If normal, suspect IC171.

If both the tape-counter and time-remaining modes are absent or abnormal, check for proper inputs to IC171. If the display is off, look for

B+ at pin 7. Then look for B+ and reset inputs at pins 20, 21, and 15. Also check for 400-kHz clock signals at pin 18. The clock signals should be about 8 V p-p. If any of the inputs to IC171 are missing, trace back to the input source.

If all inputs are present, check for normal operation of the counter and time-remaining switches by switching back and forth between the two functions. If either function is absent or abnormal, suspect the corresponding switch and/or IC171.

If all inputs and switch functions appear to be good, then look for pulses (square wave) at pin 30 of IC171 during playback. If the pulses are missing, suspect IC858 and/or IC141 (probably IC858). If the pulses are present at pin 30, but there is no counter function, suspect IC171.

13-4.3 Time-remaining operation

When IC171 is placed in the time-remaining mode by pressing the time-remaining switch (pin 23), four inputs are required for IC171 to calculate the correct time: take-up reel rotation at pin 30, supply reel rotation at pin 31, capstan servo speed for LV at pin 4, and capstan servo speed for SLP at pin 5. Without these signals, IC171 never leaves the calculation phase of the time-remaining function. With the signals, IC171 calculates the type of cassette and then changes from the calculation phase to the display phase of operation. The time remaining is then displayed and updated in all modes but special effects.

Note that when the VCR is in the special-effects modes of field still (and being frame-advanced) or slow motion, the time-remaining calculation becomes very erratic and inaccurate, so IC171 is inhibited from counting the reel pulses. This is done by a count-stop input at pin 27, supplied by a decoder within IC811.

The calculation phase of time-remaining operation is entered every time a cassette is loaded and the VCR is set to play. The cassette-down switch applies a high to pin 8 of IC171 to indicate that the cassette is loaded. IC811 applies a high to pin 28 to indicate that the VCR is in play or record mode.

Note that if the VCR is first placed in rewind or fast forward, the calculation mode is not entered, and only the tape movement direction is displayed.

The calculation mode takes approximately 20 to 60 seconds for the time-remaining calculation to conclude before the time remaining is displayed. After the first record or play function has been actuated (pins 8 and 28 high), IC171 keeps track of the time remaining until the cassette is ejected. When ejection occurs, the cassette-down switch applies a low to pin 28, instructing IC171 to clear all registers and memory.

As discussed, IC171 cannot calculate the time remaining properly when the thin eight-hour tape (such as RCA VK330) is used. In this case, the user presses the eight-hour switch (pin 3) to tell IC171 that a thin eight-hour tape is installed.

Time remaining troubleshooting/repair. If a malfunction exists in the time-remaining mode, but the counter mode operates normally, you can assume that the clock and B+reset system are good. Also, since the take-up reel-sensor input at pin 30 is used in both counter and time remaining, the pulses at pin 30 are probably good. As a result, the first place to check is at the supply-reel sensor input, pin 31 of IC171. If these pulses are missing, suspect IC858 and/or IC142. If pulses are present at both pins 30 and 31, suspect IC171.

If a malfunction exists in both counter and time-remaining modes, check for all inputs to IC171 (as discussed in this section and in Sec. 13-4.2). If any of the inputs are missing, trace back to the source. If the inputs are present, suspect IC171. Also check the display, using the general guidelines described in Sec. 13-3.5.

13-5 FINE EDITING

At the beginning of the recording process, immediately after loading, specific record voltages are developed and applied to the signal-processing circuits. Vertical sync from the incoming signal is available for drum and capstan speed/phase control, as well as to produce the CTL pulses necessary during playback. Unless the capstan motor has reached a stable running speed, transporting the tape past the CTL recording head at a fixed rate, the initial CTL pulses recorded on tape may be erratic in terms of spacing between pulses (due to uneven capstan motor speed). Also, the capstan may not be synchronized precisely with the drum until both have reached a stable speed and spacing between CTL pulses is even.

Until such time that the drum and capstan motors are synchronized to their respective reference signals and synchronized to each other, a very brief period of instability occurs. When a tape recorded under these conditions is played back, the instability is apparent at the beginning of the newly recorded information in the form of noise and unstable speed. This continues until the briefly erratic CTL pulses stabilize (even spacing) and synchronization occurs.

A more annoying problem occurs when the pause mode is used during record. Even though tape movement is stopped and the motors continue to function at full running speed, the very process of stopping the tape creates a slight movement of the tape. Because of this slight movement, the period of time is inconsistent between the last recorded CTL pulse and the first CTL pulse recorded when tape movement begins at the completion of pause. This produces uneven spacing between the CTL pulses and corresponding insta-

304 Special Features and Circuits

bility when this point is reached during playback of the tape. A similar situation can occur when pause is used on a previously recorded tape.

Some VCRs include a fine-editing function to overcome these problems. Figure 13-14 is the fine-editing circuit for a typical VCR (that shown in Fig. 1-1a). The fine-editing circuit *reverses the tape slightly at the beginning* of a record or pause mode, allowing the capstan motor to synchronize with *existing CTL pulses,* prior to allowing continuation with record. With fine-edit, there is no interruption in CTL pulse spacing. This results in a very clean, noise-free transition from pause to record, or at the beginning of newly recorded segments of a previously recorded tape.

When the record mode is selected on the VCR with the Fig. 13-14 circuit, and loading is complete, a brief reverse-search mode is initiated for a time equal to 33 CTL pulses. At that point, reverse-search is terminated and play mode is turned on for a period of two seconds. The capstan motor synchronizes to the existing CTL pulses (if any) during this period. Recording begins at the end of the two-second period. A similar action occurs when pause is selected, but with one exception. In pause, the tape remains stopped following the reverse-search segment of fine-edit. The play segment does not begin until the pause function is released.

When the record mode is initially engaged, the signal processing circuits are in the record mode. However, the SRV-REC (servo-record) voltage at pin 18 of IC5A1 remains low. Under these conditions, the servo circuits are in the play mode, responding to reference signals normally available only in playback.

The fine-edit input at pin 4 of IC5A1 is normally high. This maintains a high at pin 14 of IC5A0 and charges C5A2 to a high through R5B6. When record is first selected, IC5A1 outputs a high at pin 22, placing the VCR in reverse-search mode. Simultaneously, the high at pin 4 of IC5A1 is removed. However, the high at pin 14 of IC5A0 remains, due to the charge on C5A2.

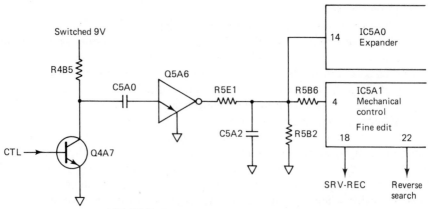

FIGURE 13-14 Fine-editing circuits

During the reverse-search segment of fine-edit, if no CTL pulses are present on the tape in use, C5A2 discharges toward a low through R5B2. When the low reaches the threshold level at pin 14 of IC5A0, the reverse-search mode is terminated through action of IC5A1.

If CTL pulses are present on the tape, C5A2 is discharged at a faster rate. During the reverse-search segment, CTL pulses are amplified by Q4A7 and differentiated by C5A0/R4B5. The positive-going pulse resulting from this differentiation momentarily turns on Q5A6. This connects R5B2 in parallel with R5E1, reducing the resistance to ground in the discharge path of C5A2 and permitting C5A2 to discharge faster. The component values are chosen so that the charge across C5A2 reaches a low following the 33rd CTL pulse. This low terminates the reverse-search segment and initiates the play segment of fine edit.

During the play segment, the SRV-REC output at pin 18 of IC5A1 remains low, maintaining the servo circuits in playback. The VCR remains in play for about two seconds, during which time the servo circuits are allowed time to synchronize to any previously recorded CTL pulses. At the end of the two-second period, IC5A1 automatically terminates the play mode and drives the SRV-REC high, placing the VCR in the normal record mode.

When pause is first selected, the reverse-search mode is also engaged for a time period equal to 33 CTL pulses. The reverse-search segment is then terminated as described. However, IC5A1 is programmed so that the play mode does not follow reverse-search during pause. Instead, play does not begin until pause is released. At that time, IC5A1 issues play commands for a period of two seconds, followed by a high SRV-REC output. This places the VCR in a normal recording mode and results in a very smooth transition of recorded material at the points on tape where the pause mode is activated.

Fine-edit troubleshooting/repair. The classic symptom of a malfunction in the fine-edit circuit is excessive noise or picture instability when playback first starts (first two or three seconds), or when pause mode is selected. The problem is usually easy to identify, and the circuits can be localized quickly. (It is often the timing capacitor that causes the problem.) However, keep two points in mind when troubleshooting the fine-edit circuits: Not all VCRs have fine-edit circuits, or any comparable circuit. If the VCR does have a fine-edit circuit, but you are playing a tape recorded on a VCR without fine-edit, you will get the same symptom.

13-6 INDEX OPERATION

Most VCRs have some form of index operation which permits the user to locate quickly specific segments of a recorded tape. The function is called by various names, such as *index search, memory index, automatic index,* or simply *index operation*. The most common system uses the electronic counter as described in Sec. 13-3.1. You set the electronic counter to 0000 when you reach the beginning of the desired tape segment. When you want to return to that segment, you press a front-panel memory button. The VCR goes into rewind and stops when the counter reaches 0000. You can start playback at the beginning of the segment, without further search on your part.

Some modern VCRs have an improved index operation using an additional head. The basic arrangement is shown in Fig. 13-15. Note that both the full-erase head and special index head are required. During record, the VCR is placed in the stop mode between recorded selections. When the index feature is selected, a 30-Hz signal is recorded across the width of the tape by the full-erase head. The 30-Hz signal is then read by the index head during fast forward of the tape. When a 30-Hz index signal is picked up by the index head, the VCR is stopped.

13-6.1 Typical index operation circuits

Figures 13-16 and 13-17 show the index-record and index-playback circuits, respectively, for a typical VCR (that shown in Fig. 1-1f). Note that a separate

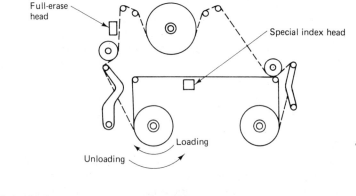

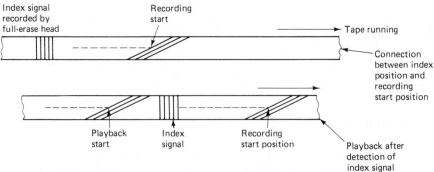

FIGURE 13-15 Basic arrangement for full-erase and special-index heads

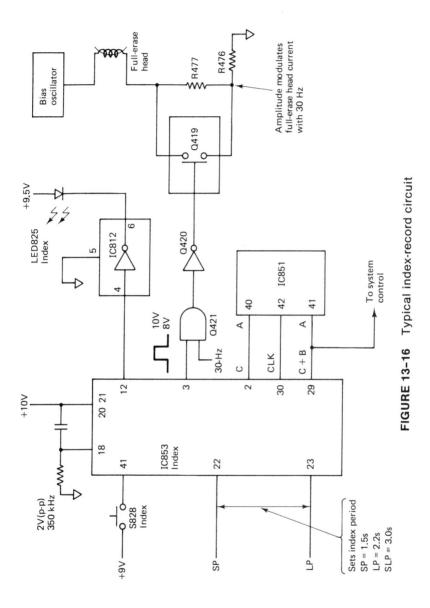

FIGURE 13-16 Typical index-record circuit

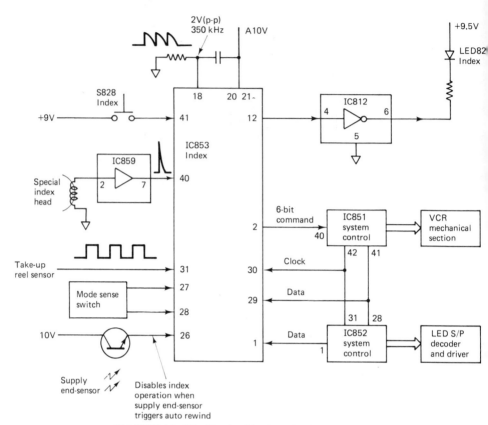

FIGURE 13-17 Typical index-playback circuit

microprocessor IC853 is used to perform index operations. During record, IC853 (when turned on) instructs the full-erase head circuits to apply an index signal to the tape. The index signal is recorded during the period when the VCR is placed in record from the stop mode. With the index mark recorded at the beginning of each recorded segment, if the VCR is placed in fast forward or reverse, the circuits monitoring the tape detect the index signal.

The detected signal is applied to IC853, which then instructs system control to place the VCR in forward search. The forward-search mode is continued for 10 seconds, allowing the user to enter a different command, if desired. If after 10 seconds no other function has been selected, IC853 returns the VCR to the previous mode (either rewind or fast forward) and continues looking for any additional index signals.

13-6.2 Index-record operation

As shown in Fig. 13-16, index record is selected when the index switch S828 is pressed. This applies a high to pin 41 of IC853, which responds by outputting a high at pin 12. This high is inverted by IC812 and turns on front-

Index Operation 309

panel index LED825 (indicating that the index function is in operation). Operating-mode information from system control IC851 is applied to IC853 at pins 29 and 30. When the mode changes from stop to record, IC853 outputs a high at pin 3. The duration of this high is determined by the tape-speed logic signals at pins 22 and 23 of IC853 (SP = 1.5 s, LP = 2.2 s, SLP = 3.0 s duration).

The index high at pin 3 of IC853 is passed to the enabling input of AND-gate Q421. The other input to Q421 is from the switched 30-Hz signal taken from the cylinder servo (Chapter 7). During that time when the high appears at pin 3, the 30-Hz signal is passed through Q421 and applied to Q420 where the signal is inverted. The 30-Hz signal is then applied to Q419, which acts as an on-off switch shorting across R477. The 30-Hz signal causes the resistance of R477 to be reduced to zero, and then back to full resistance, at a 30-Hz rate. This continues for the same time period that the output at pin 3 of IC853 remains high.

The action of R477 going between full resistance and zero resistance modulates the full-erase head current at a 30-Hz rate. Due to the strength of this amplitude-modulated full-erase signal, the erase field impregnates the surface of the tape well into the tape to produce a strong index recording. This is necessary so the video heads that apply the recorded video track on tape do not disturb the index signal.

Index-record troubleshooting/repair. Before you troubleshoot the index circuits, first check for normal operation in a variety of operational modes (play, record, pause, slow motion, etc.). If you get normal operation in all modes, then the system-control circuits and system-control portions of IC853 are functioning normally. If so, look for a 2-V pulse (from a steady 8 V to about 10 V) at pin 3 of IC853 when index switch S828 is pressed and record is selected (when going from stop to record). If the pulse is missing, suspect IC853.

Next look for the gated 30-Hz switching signal at the output of Q420. If missing, suspect Q421 and Q420. If present, check for a modulated erase current at the full-erase head. (This can be done by measuring the voltage across R476.) If modulation is absent (no voltage or a steady voltage across R476), suspect Q419. If the 30-Hz modulation is present, then the index-record operation is functioning normally. Check the index-playback operation.

13-6.3 Index playback operation

As shown in Fig. 13-17, index playback is selected when the index switch S828 is pressed. IC853 then monitors for an index signal at pin 40, but only during rewind or fast-forward modes. The mode information to instruct IC853 when the VCR is in rewind or fast forward is obtained from system control microprocessors IC851 and IC852.

When indexing is turned on during fast forward, the index signal is recovered from the index head and applied to pin 40 of IC853 through amplifier/detector IC859. When IC853 receives the index pulse, a 6-bit command is applied to IC851 from pin 2 of IC853. This command places the VCR in forward search. IC853 maintains forward search for approximately 10 seconds, or until the user selects another mode. After the 10-second period, if no other mode is selected, IC853 outputs a 6-bit command, which causes IC851 to place the VCR back in fast forward. The VCR then continues in the fast-forward mode until the next index mark occurs.

This process continues to the end of the cassette tape unless the user selects another mode. When the tape reaches the end of tape, the supply end sensor triggers the auto rewind mode. The tape end sensor also applies a high to pin 26 of IC853 to indicate that auto rewind is activated. This prevents IC853 from stopping the tape in the auto-rewind mode at the index marks.

If indexing is turned on during rewind, when the index mark triggers the forward search mode, the amount of tape transported during the forward search is monitored by IC853. The reel-rotation signal at pin 31 (from the take-up reel sensor) is applied to a counter within IC853, incrementing the counter. When forward search is completed, the VCR transfers back to the rewind mode. The counter is switched over to the decrement (subtract) mode so that any index signal played back (within two seconds after the internal counter is set to zero) does not turn on the forward-search mode.

Index playback troubleshooting/repair. First confirm that the proper index signal is recorded on tape as described in Sec. 13-6.2. Select fast forward and check for the presence of an index pulse at pin 40 of IC853. If known good signal is on the tape, and the pulse is missing at pin 40, suspect IC859 or the index head. If the pulse is present at pin 40, but the VCR does not go into forward search from fast forward, suspect IC853.

After a recording period when indexing has been selected, if the end of tape is reached and the VCR enters the auto-rewind mode and the mechanism stops on all index signals in auto rewind, suspect IC853.

14

SYSTEM CONTROL

This chapter is devoted to troubleshooting and repair of VCR system-control circuits. System control is an area where one model of VCR can differ greatly from other models. However, operation of the system-control circuit for any VCR is determined primarily by microprocessors. Most modern VCRs use several VCRs in system control. The microprocessors accept logic-control signals from the VCR operating controls (typically feather-touch push buttons) and from various tape sensors. In turn, the microprocessors send control signals to video, audio, servo, and power supply circuits, as well as drive signals to solenoids and motors.

We do not go into operation of microprocessors here since such information is beyond the scope of this book (and each VCR has its own particular microprocessor applications). However, we do discuss the circuit inputs and outputs to and from the microprocessor, since such circuits are typical for many VCRs. If you want a thorough discussion of microprocessors, your attention is invited to my best-selling *Handbook of Microprocessors, Microcomputers, and Minicomputers* (Englewood Cliffs, N.J.: Prentice-Hall, Inc., 1979).

Note that many system-control functions are closely related to mechanical operation of the VCR, as well as to many circuit functions. So it is essential that you study the related chapter when reviewing system-control operation. In most cases there is a reference from the related chapter to this chapter when system control is involved.

Note that some VCR service literature uses the term *microprocessor,* while other literature favors the term *microcomputer.* In both cases, we are

312 System Control

discussing a single IC that performs many specific, computerlike functions in response to various inputs. I prefer the term *microprocessor,* and use the term throughout this chapter. But do not be surprised to find a VCR described as having many microcomputers in its system-control circuits.

14–1 MICROPROCESSOR COMMUNICATIONS BUS

Most modern VCRs have more than one microprocessor in system control. This requires a means of communications among the microprocessors. Typically, some form of communications bus is used. The communications bus is similar to those used in microprocessor-based computers.

Figure 14–1 shows the microcomputer communications bus system for a typical VCR (that shown in Fig. 1–1c). As shown, the VCR has a total of five microprocessors, which communicate to each other through a variety of parallel and serial communications buses. Within the tuner/timer/charger there are three microprocessors: remote-control IC401 (Chapter 4), tuning/PLL IC701 (Chapter 10), and timer IC101 (Chapter 13). Note that remote IC401 is the master microprocessor in this group. IC401 and IC101 pass channel-selection data to IC701 on a *4-bit parallel data bus. Handshaking* control lines are used to maintain communications control, as discussed in Sec. 10–3 of Chapter 10.

Two microprocessors IC801 and IC802 perform system-control functions. IC801, designated as microprocessor-A, is considered the master of

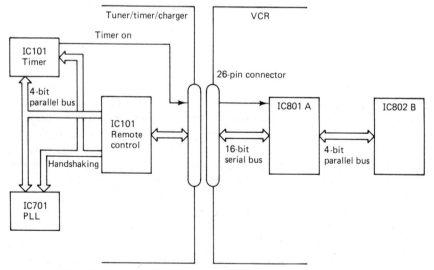

FIGURE 14–1 Typical microcomputer/microprocessor communications bus

IC802, designated microprocessor-B. IC801 and IC802 communicate to each other on a *4-bit bi-directional parallel bus*. In turn, IC801 communicates with master remote IC401 on a *16-bit serial data bus* (routed through a 26-pin connector located at the rear of the VCR). Data bits are passed between IC401 and IC801 during the time that either the IR remote or the front panel is turned on. The 16 serial data bits, along with handshaking signals, are transferred on the 16-bit bus.

Throughout this chapter we discuss troubleshooting procedures and methods to help isolate which microprocessor is not operating properly. However, the conventional troubleshooting procedures in servicing radio, TV, stereo, and the remainder of VCR circuits cannot be applied to servicing digital-data communications systems using communications buses. A conventional oscilloscope (even a storage scope) is not of much value when trying to monitor the various complex signal patterns passed between microprocessors. The *logic analyzer* is your best bet to do this job of monitoring. Unfortunately, most TV/stereo shops do not have a logic analyzer (discussed in Sec. 2-2.6). So we describe *alternate methods to get around the problems of data-communications troubleshooting,* using less exotic (and less expensive) test equipment. Before we get into these methods, let us see how the microprocessor-based system-control functions are performed.

If you are really interested in digital data communications systems, including communications buses and test equipment, read my *Handbook of Data Communications* (Englewood Cliffs, N.J.: Prentice-Hall, Inc., 1984).

14-2 SYSTEM-CONTROL OVERVIEW

Figure 14-2 is the block diagram for the system-control circuits (of the VCR shown in Fig. 1-1c). Microprocessor-A IC801 receives the various input commands from the tuner/remote IC (through the 16-bit serial data bus), the camera remote input, and/or the front-panel operation buttons. IC801 also controls the power on/off circuits (Chapter 3), which latch the power relay. IC801 communicates to microprocessor-B IC802 (through a 4-bit data bus).

Microprocessor-B IC802 drives the mechanical systems (Chapter 5) of the VCR deck and determines the condition of the deck (tape loading, cassette loading, etc.) in all modes of operation. IC802 also drives the LCD counter display module (Sec. 14-6) and monitors the various trouble sensors (Sec. 14-5), which protect the system should a malfunction occur.

IC801 and IC802 function together as one microprocessor (and are physically mounted on the same system-control board in this VCR). As discussed, it is very difficult (near impossible) to monitor the data communications between IC801 and IC802. So the manufacturer generally recommends that both microprocessors be replaced as a package, along with the system-control board.

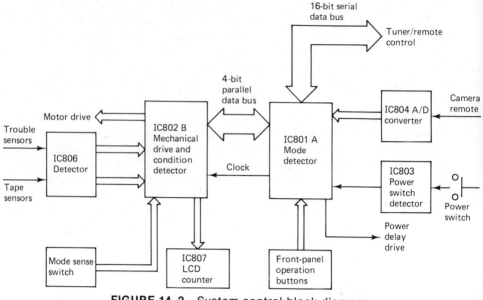

FIGURE 14-2 System-control block diagram

This is a practical approach since both ICs are flat-pack types and very difficult to replace on an individual basis.

Even though the microprocessors are replaced as a package, it is still helpful in troubleshooting to know what inputs are applied to which microprocessor, and how both microprocessors produce outputs in response to these inputs. The following paragraphs summarize these functions.

14-2.1 System control microprocessor-A IC801 inputs/outputs

Figure 14-3 shows the functional sections of microprocessor-A. As shown, IC801 is divided into eight functional sections.

Pins 25, 34, and 35 are used during the on/off function of the VCR (Sec. 14-3). The power-hold IC803 is connected to the on/off switch. When the switch is pressed, a low is applied to pin 25. This signals IC801 that the on/off command is requested and that if the VCR is in the off mode, IC801 is to apply power and turn on the VCR. If the VCR is already on, the information is interpreted to turn the VCR off. This is done by IC801 generating output phase-0 ($\emptyset$ 0) and phase-1 ($\emptyset$ 1) signals at pins 34 and 35, respectively.

Pins 19, 20, 21, and 22 become active with application of B+ power. When B+ is applied, reset occurs and the 3.58-MHz oscillator begins operating.

The cassette safety tab (Sec. 1-9) is connected to pin 16 of IC801. If the

System-Control Overview 315

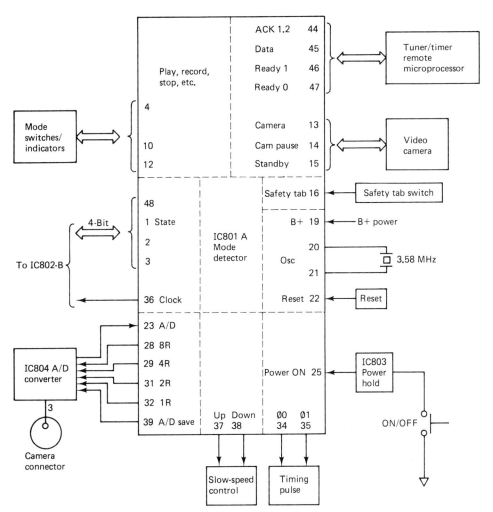

FIGURE 14-3 Functional sections of system-control microprocessor-A

tab is removed, IC801 recognizes this condition and prevents the VCR from going into the record mode.

Pins 13, 14, and 15 of IC801 monitor the external video camera (Sec. 14-4) and determine when a camera is connected. The camera can request pause and camera standby through these pins. Pins 44, 45, 46, and 47 provide communication to the tuner/remote control.

Pins 4 through 12 are the multiplexed input/output pins used to accept signals from the mode switches (front-panel operation buttons), and to drive the front-panel LED indicators.

316   System Control

Communication between IC801 and IC802 is done over the 4-bit parallel bus (pins 1, 2, 3, 48) and a clock line (pin 36). When an input command is applied to IC801, the command is relayed to IC802 over the bus. When a VCR function such as automatic stop, end of tape, or trouble-sensor indication is monitored by IC802, a stop command is passed to IC801 over the bus.

The various mode functions (play, rewind, etc.) which can be controlled by the external camera are interpreted by A/D (analog/digital) converter IC804 through pins 23, 28, 29, 31, 32, and 39 of IC801.

Slow-speed information received from the remote-control IC401 is output from pins 37 and 38 of IC801 to the slow-speed control circuits.

14-2.2 System-control microprocessor-B IC802 inputs/outputs

Figure 14-4 shows the functional sections of microprocessor-B. As shown, IC802 is divided into many functional sections. IC802 responds to mode commands from IC801 and passes the information to the various mechanisms within the VCR deck. Also, IC802 monitors the VCR deck mechanical mode sense switches and troubleshooting sensor, informing IC801 when a malfunction occurs. The functional sections of IC802 are as follows:

IC802 communicates with IC801 on the 4-bit parallel data bus (pins 38, 39, 40, and 41) and clock line (pin 37).

Various trouble sensors are input at pins 3 through 6, 28 and 29.

When the tuner/timer/charger is programmed for a timed recording (Chapter 13), and the appropriate time has occurred, the information is transferred from the timer IC101 to pin 30 of IC802. This instructs IC802 that a time recording has occurred and to place the VCR in record mode.

The battery voltage is monitored at pin 31 of IC802. When battery voltage falls below 10.5 V, the VCR is placed in stop mode.

The mechanical position or operational mode of the VCR is monitored at pins 33, 34, and 35 of IC802.

The servo circuit is controlled by a variety of outputs from IC802 at pins 7, 10, 12, and 16 through 20.

The various B+ signals found within the VCR are developed from the outputs at pins 13, 14, and 15 through mode-switching circuits.

The audio circuits are controlled by pins 8 and 9 during audio dub mode.

The LCD display IC807 is controlled by signals at pins 42 through 47. The LCD display is a self-contained digital counter and display driver on one modular circuit board (in this VCR). The assembly is nonserviceable and must be replaced as a module or package. (This is typical for many modern VCRs.)

The power-save function (which is required during portable operation) is contained within IC802. During camera operation, if the VCR is left in stop or pause for five minutes, the power-save line goes high. This turns off the

System-Control Overview 317

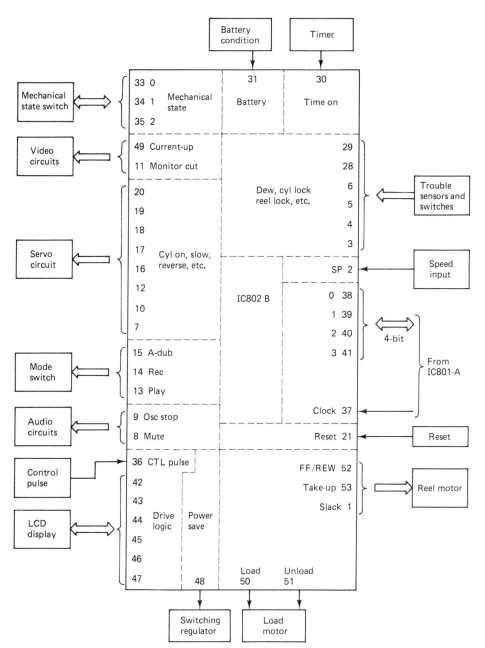

FIGURE 14-4 Functional sections of system-control microprocessor-B

318 System Control

switching regulator (Chapter 3), placing the VCR in power-save mode to conserve battery power.

Control of both loading and reel motors (Chapter 5) is done at pins 1 and 50 through 53.

14-3 VCR ON/OFF OPERATION

Figure 14-5 illustrates the on/off circuits (for the VCR shown in Fig. 1-1c). On/off operation is done by controlling relay RL701. When the front-panel power on/off switch S809 is pressed, or a command is sent from remote-control IC401, a low is applied through diode D810. This low pulls pin 16 of IC803 and pin 2 of IC704 to ground.

When pin 2 of IC704 is grounded, IC704 energizes relay RL701. This causes the +12 V from the power supply or battery to be applied to the input of the switching regulator, which then outputs various dc voltages (Chapter 3).

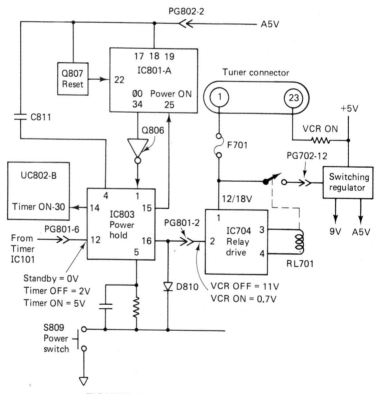

FIGURE 14-5 VCR on/off circuits

One voltage output from the switching regulator is the A5 voltage. The A signifies that the voltage is always present when the VCR is turned on. The A5-V output supplies power to IC801, and the reset circuit of Q807. A5 V also charges capacitor C811, generating a momentary pulse at pin 4 of power-hold IC803. The pulse sets an internal latch, causing pin 15 of IC803 to go low. If the voltage at pin 12 of IC803 is not 2 V (timer off), the latch circuit does not generate a low at pin 15.

The low at pin 15 of IC803 is input to pin 25 of IC801, the power-on input of microprocessor-A. With B+ (pin 18), reset (pin 22) and a low at pin 25, IC801 generates phase-0 pulses at pin 34. The phase-0 pulses are inverted by Q806 and applied to pin 1 of power-hold IC803.

If inverted phase-0 pulses are present at pin 1 of IC803, indicating that power has been turned on, IC803 latches RL701 (through IC704) in the on mode. This is done by latch circuit within IC803 that pulls pin 16 low. The low is maintained at pin 2 of IC704 for the duration of the VCR on time.

When switch S809 is pushed again (to turn off the VCR) or an off command is received from the remote IC401, pin 5 of IC803 is pulled low. With pin 5 low, IC803 resets the internal latch, allowing pin 16 to change to a high. This high is applied to pin 2 of IC704, which then deenergizes RL701 and disconnects the 12-V power from the switching regulator.

During the power-on mode, IC803 monitors the timer-on line at pin 12. One of three different voltages appears at pin 12, as shown in Fig. 14-5. If the timer has not been preset for a timed recording, and the timer switch is in the off position, a 2-V level appears at pin 12. If a programmed time is input to the timer microprocessor IC101, and the timer switch is turned to the on position, the time-on line at pin 12 is zero volts. This signals IC803 not to allow manual on/off operation of the VCR.

The third voltage at pin 12 of IC803 occurs when the timer is turned on at the start of a timed recording, and the line at pin 12 is pulled to 5 V. This voltage signals IC803 to pull pin 15 low. IC801 then generates phase-0 pulses and IC803 goes low, turning the VCR on. At the same time, IC803 also sends a time-on signal to IC802 at pin 30. This signal instructs IC802 to place the VCR in record.

When the end or off-time of the timed recording occurs, the voltage at pin 12 of IC802 returns to the 2-V level, IC803 pulls pin 16 high, deenergizing RL701 and turning off the power-supply circuit to the VCR deck.

VCR on/off circuit troubleshooting/repair. If you cannot turn the VCR on (or off), the first step is to monitor inputs and outputs on the system-control board. Keep in mind that the two microprocessors IC801/IC802 are not serviceable, and that the manufacturer recommends replacement of the entire system-control board in the event of malfunction. However, before you order a new system-control board (at mind-boggling expense to the customer), make the following checks.

Check for 2-V power at pin 6 of plug PG801. If incorrect, (carefully) unplug the system-control board, and check for 2 V at pin 6 of socket CN701 (which mates with PG801). If you do not get 2 V, the problem is likely in the timer circuits (Chapter 13) and not system control. Fortunately, many of the timer circuits for most modern VCRs are replaceable. Of course, if 2-V power is available at pin 12 of IC803, the system-control board is probably at fault.

If the voltage at pin 6 of PG801 is correct, then confirm that the voltage level at PC801-2 is 11 V (when the power switch S809 is not pushed) but goes to about 0.7 V (when S809 is pushed and held in). If not, suspect S809 and the system-control board.

Next, look for +12 V at PG702-12. If missing, suspect F701, IC704, or RL701 (all of which are replaceable). Then look for +5 V at PG802-2. If missing, suspect a defective switching-regulator module (Chapter 3).

If the voltage at pin 12 of PG702 is present when S809 is pressed, but does not remain when S809 is released, suspect the system-control board (and the customer must pay!).

14-4 SYSTEM-CONTROL INPUT OPERATION

Figure 14-6 illustrates the system-control input circuits (for the VCR shown in Fig. 1-1c). The VCR operation mode is selected from three different sources: the front-panel keyboard, the tuner/timer remote control IC401, or the camera input connector. Microprocessor-A IC801 monitors these three input sources.

Keyboard operation is done by IC801 generating phase-1 scanning pulses at pin 35. These pulses are amplified by Q805 and applied to the keyboard switches. IC801 also generates phase-0 pulses, which are output at pin 34 and amplified by Q806 to drive the LED displays on the front panel. The return signals from the keyboard and LED displays are routed to eight key-in/display-drive input pins on IC801.

IC801 multiplexes the function of keyboard input and display drive. During the time when phase-0 is active, LED drive operation is being performed. When phase-1 is active, keyboard scanning occurs, and IC801 monitors for a command or function input.

IC801 also monitors for input from remote-control IC401 (over the 16-bit serial data bus and handshaking lines) at pins 44 through 47. The ready-1 (pin 46) and ready-2 (pin 47) lines inform the microprocessors that one or the other is ready to send data. After receipt of the ready signal from the transmitting microprocessor, the receiving microprocessor outputs an acknowledge signal on the acknowledge-1,2 line. The transmitting microprocessor then acknowledges this signal and outputs serial data on the data line (pin 45). This sequence occurs for each 16-bit word.

The third source of input is applied through pin 3 of the external camera

System-Control Input Operation 321

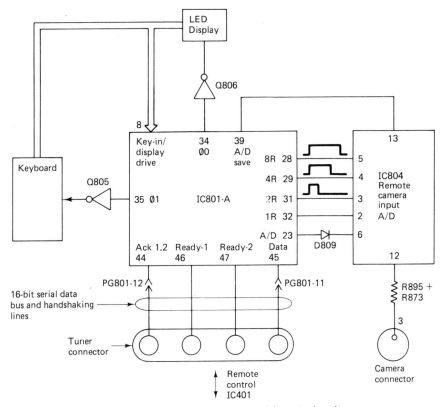

FIGURE 14-6 System-control input circuits

connector to IC804 (the camera input A/D converter). IC801 passes four scanning pulses (pins 28, 29, 31, 32) to the IC804 (pins 2, 3, 4, 5) and monitors the A/D signal at pin 23 (applied through D809). The camera circuit connects a predetermined value of resistance from pin 3 to ground for the mode required. A constant-current source within IC804 is passed through the camera resistance, developing a voltage at pin 3 proportional to the resistance. IC804 compares this voltage to the voltage developed from the combination of 8R, 4R, 2R, and 1R outputs from IC801.

IC801 monitors the signal at pin 23 for an indication when the voltage developed from the 4-bit signals matches the voltage developed from the external resistor (at pin 3 of the camera connector). The 4-bit signal is then decoded by IC801 to determine the function requested.

You can see from this that simply monitoring the levels on the 4-bit bus with a scope (or even a probe) will be of little value in troubleshooting. You need a logic analyzer to monitor the 4-bit coding between IC801 and IC804. It is equally difficult to monitor the 16-bit serial coding between IC801 and IC401. However, there is a way around this troubleshooting problem.

System-control input troubleshooting/repair. If you get good remote-control operation, but no manual operation, the problem is likely in the keyboard circuits or Q805. (In this particular VCR, you must replace the entire system-control board, increasing your profits, in that event.)

If the opposite is true, and you have good manual operation but no remote, look for an indication of signal activity on the ready-1 and ready-2 inputs of IC801 (at pins 7 and 13 of PG801). You can do this with an oscilloscope or logic probe. (You should see evidence of rapidly changing pulses on both ready lines, if normal communications is being passed between IC801 and IC401.) If pulse activity is missing, suspect remote-control IC401, the remote-control circuitry, or the hand-held IR remote unit (all of which are discussed in Chapter 4).

If you get pulse activity on the ready lines, look for similar activity on the acknowlege-1.2 line at PG801-12. If missing, replace the system-control board.

If you get pulse activity on the acknowledge-1.2 line, look for similar activity on the data line at PG801-11. If missing, suspect the remote control IC401 (or IR remote unit).

If you get pulse activity on the data line, try substituting the system-control board. Of course, it is possible that either IC401 or the IR remote unit is producing some unintelligible code which cannot be interpreted by IC801. However, this is a long shot.

If you get good manual and remote operation, but no camera function control, try a different camera. If this is not practical, try substituting various resistors from pin 3 of the camera connector to ground. The resistor values for the various modes are listed in Fig. 14-7. For example, if a 15.4-k resistance is connected between pin 3 of the camera connector and ground, the VCR should go into slow motion.

Mode	Camera connector-pin 3-to-ground resistance (kΩ)
Play	111
Record	51.4
Frame advance	21.4
Slow	15.4
Search forward	8.5
Search reverse	6.4
Fast forward	4.7
Rewind	3.4
Stop	2.3
Pause	0

FIGURE 14-7 Camera operating mode versus input resistance

If you do not get the correct mode with a resistor connected at pin 3, suspect IC804 or IC801 (in that order). Either way, you must replace the entire system-control board.

If you do get the desired mode of operation when the correct resistance is connected at pin 3, look for problems in the camera cable or in the camera circuits (if you feel courageous enough to tackle a video camera repair job!).

14-5 TROUBLE SENSORS

Figure 14-8 illustrates the system-control trouble sensor circuits (for the VCR shown in Fig. 1-1c). IC802 monitors the various trouble sensors and detectors: dew sensor, rewind or forward, tape-end, reel-rotation, cassette-up, and cylinder-lock detectors.

14-5.1 Dew sensor

The dew sensor element is connected to plug PG805, pin 6. If there is low moisture within the VCR, the resistance of the dew sensor is high. This produces a high at pin 9 of IC809, a high at output pin 14 of IC806, and a high at pin 6 of IC802, which responds by doing nothing. However, if there is high moisture within the VCR, the resistance of the dew sensor drops, producing a low at pin 6 of IC802, and the VCR is placed in the stop mode.

14-5.2 Rewind or forward tape-end detectors

The rewind tape-end detector Q903 senses the *infrared signal* that is passed through the clear leader portion of the VHS tape when the end of tape is reached. This generates a high at pin 5 of IC806 and pin 28 of IC802, which responds by placing the VCR in stop.

Note that most modern VHS VCRs use *infrared end-of-tape sensors and light sources.* This is not true of early-model VCRs, which use ordinary light and photodetectors. The use of conventional light can be a problem during service. If you remove the tape, or remove some covers that expose the sensors to outside light, the VCR goes into stop, ready or not. So you must cap the sensors. As discussed in Sec. 2-7.6, do not remove the sensor lamp or light source. On many VCRs this triggers the VCR into the stop mode. These problems do not exist on Beta VCRs, which use a metal foil at the end of tape.

The forward end-of-tape detector Q904 senses the infrared signal passed through the clear leader at the forward end of tape. This generates a high at pin 7 of IC806 and pin 29 of IC802, which responds by placing the VCR in stop.

324 System Control

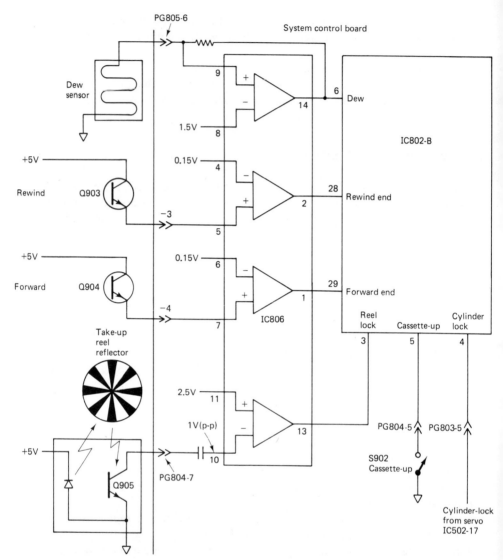

FIGURE 14-8 System-control trouble sensor circuits

14-5.3 Reel-rotation detector

Detection of take-up reel rotation is done by a combined LED and light-sensor transistor in one package, Q905, which is located under the take-up supply turntable. A reflector disc with eight reflective and nonreflective areas is located on the bottom of the take-up turntable. As the take-up reel rotates, square-wave pulses are generated by Q905 (as described in Sec. 13-3.2) and applied to the system-control board at pin 7 of PG804.

The pulses are applied to pin 3 of IC802 through IC806 and a capacitor. If the reel stops rotating, the square-wave pulses are removed and IC802 places the VCR in stop. During normal operation, while the take-up reel is rotating, the pulses are divided in IC802 and used to feed the LCD tape counter (Chapter 13).

14-5.4 Cassette-up switch

The cassette-up switch S902 is connected to pin 5 of IC802. When a cassette is loaded into the holder, and the holder is moved to the down position (Chapter 5), a high is applied to pin 5 of IC802. If the cassette-up switch opens (even momentarily) for any reason during normal operation (play, rewind, slow, etc.), the high is removed from pin 5 and IC802 places the VCR in stop.

14-5.5 Cylinder-lock detector

The cylinder-lock signal is supplied from the cylinder servo (Chapter 7) IC502, at pin 17, and applied to pin 4 of IC802. During normal operation, if the cylinder motor speed decreases, a high is generated by IC502. This high causes IC802 to place the VCR in stop.

Trouble sensor troubleshooting/repair. If the VCR does not go into play, check the inputs of all trouble sensors at the corresponding pin of the system-control board. Look for a voltage of less than 1.5 V at the dew sensor input PG805-6, and less than 0.15 V at end-of-tape sensors at PG805, pins 3 and 4. Also look for a high at pin 5 of PG804 from the cassette-up switch (make sure that there is a cassette in place). If any of these trouble-sensor inputs are incorrect (voltages substantially higher than 1.5 for the dew sensor, and higher than about 0.15 V for the end-of-tape sensor), track the signal back to the source (sensors or switch). If all signals are normal, replace the system-control board.

Note that if the VCR does not make any attempt to load tape from a cassette (after the cassette is pulled in and down by the front-load mechanism), the reel-rotation detector and cylinder-lock detector are *probably not at fault*. These two trouble sensors place the VCR in stop only after the cassette is loaded, the tape is loaded from the cassette, and the tape starts to move.

If the VCR loads tape from the cassette, and then immediately unloads tape and stops, look for a 1-V square wave at PG804-7 at the time when the tape just starts to move (take-up reel rotating). If the pulses are missing, suspect Q905, reel motor problems (Chapter 5), or capstan servo problems (Chapter 8). If the pulses are present, check for a cylinder-lock signal at pin 5 of PG803. The cylinder-lock line should be high while the tape is being loaded from the cassette, and then go low at completion of tape load (when the tape starts to move). If not, suspect cylinder servo problems (Chapter 7).

14-6 COUNTER/RECORD TIME

Figure 14-9 illustrates the counter/record-time circuits (for the VCR shown in Fig. 1-1c). Note that in this VCR, the counter/record-time indicator is considered part of system control rather than a separate special-feature circuit as described in Chapter 13. Also, note that the VCR in Fig. 14-9 uses a liquid crystal display (LCD) counter/record-time indicator rather than the fluorescent display described in Chapter 13. The LCD display and associated circuits shown in Fig. 14-9 are contained on a *nonserviceable display module,* located on the front of the system-control board.

The display performs three modes of operation: record-time indication, counter indication, and battery empty (battery-E).

During record-time and counter operation, the control track pulse (CTL) at pin 7 of PG803 is amplified by Q802 and applied to pin 36 of IC802, which divides the pulse by 30. The divided-down CTL pulse (now at 1 Hz) is applied to pin 13 of the display module and used to count the time in record mode.

The output from the reel-rotation detector (amplified by IC806) is applied to pin 3 of IC802, where the signal is divided by 8, producing a one-pulse-per-reel-rotation signal at pin 47 of IC802. This signal is applied to pin 14 of the display module and used as the tape-counter signal to drive the four-digit tape counter.

The circuits within the display module must be able to count both up and down, and must be able to recognize when the VCR is in record, playback, FF, or reverse. This is done by *mode control* signals applied to the display module from IC802 (at pins 43, 44, and 45). Figure 14-10 shows the relationship between the mode-control signals and the operating mode selected. For example, when all three mode-control signals are high, the count stops. When mode-0 is low, and both mode-1 and mode-2 are high, the counter counts up, and the counter memory is turned on.

During various modes of operation, the counter memory circuit is turned on to tell the VCR when a reading of 0000 is reached, and to place the VCR in stop. The memory function is activated when the display-module memory switch is pressed. This applies a 0000 signal to pin 42 of IC802 when the count reaches 0000.

The battery-E signal from Q803 is applied to pin 3 of the display module. When the battery charge falls below the serviceable level (typically 10.5 V), an alarm signal is applied to Q803. The battery-E display appears when pin 3 of the display module goes low.

B+ power for the display module is taken from a two-transistor voltage regulator circuit on the system-control board. Transistors Q804 and Q816 produce a 3-V regulated supply, which is filtered by C803 and C804. These ca-

Counter/Record Time   327

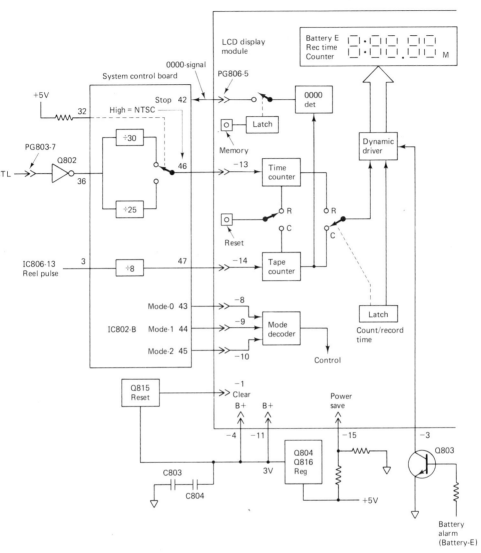

FIGURE 14-9 Counter/record-time circuits

pacitors are the *power backup* system in the event dc power is lost. C803 and C804 can store enough charge to maintain LCD display modulation for about one hour. This is done by the *power save sense* line at pin 15 of the display module. If the 5-V supply is lost, the counter is turned off, thus reducing power consumption.

Counter/record-time troubleshooting/repair. As discussed, the LCD display module is nonserviceable. So the first step in troubleshooting is to

Mode	IC802-B			Operation
	Mode-0 Pin 8	Mode-1 Pin 9	Mode-2 Pin 10	
Stop	H	H	H	Count stop
Play, forward search	L	H	H	Up-count, memory counter-effective
Reverse play, reverse search	L	L	H	Down-count, memory counter-effective
Fast-forward	H/L	H/L	L	Up-count, time indication is switched over to tape counter indication. Counter reset, memory counter-effective
Rewind	H	L	H	Down-count, time indication is switched over to tape counter indication. Counter reset, memory counter-effective

FIGURE 14-10 Counter/record-time mode control chart

check inputs to the display. If any input is absent or abnormal, trace the line back to the source. If all the inputs are present and appear to be normal, replace the display module. For example:

If there is no display in any mode, check the power inputs at pins 4, 11, and 15, and the reset input at pin 1. Also look for shorted C803/C804 or defective Q804/Q816.

If there is no battery empty (battery-E) display, check the input at pin 3 and Q803.

If there is no tape-counter mode, but record time is good, check the input at pin 14.

If there is no record-time mode, but tape-counter mode is good, check the input at pin 13.

If the display appears to be operative in some modes, but not in one or more modes, check the mode-control signal inputs at pins 8, 9, and 10 using Fig. 14-10. For example, if the counter counts up (during play or forward search) but does not count down (during reverse play or reverse search), check that mode-0 and mode-1 (pins 8 and 9, respectively) are low, and that mode-2 (pin 10) is high. If the mode-control signals are good, replace the display. If the mode signals are not good, replace the system-control board.

If the display goes to 0000, but the VCR does not stop when the memory button is pressed, check for an output from pin 5 of display module connector PG806 to pin 42 of IC802 on the system-control board. This output, the 0000 signal, should be present when the count is 0000 and the memory button is pressed. If not, replace the display module. If the 0000 signal is present at IC802, but the VCR does not stop, replace the system-control board.

INDEX

A

AFC compatability problems, 79
AFT channel scan, 230
Alignment tapes, 69
Alternations, design, 63
APC circuits (VHS), 39
Audio:
 circuits, 251
 dubbing, 5, 52, 244
Azimuth loss recording, 16, 30

B

Background noise, 245
Bandswitching, 229
Battery charge circuits, 96
Beta:
 circuits, 254
 PI color recording, 19
 playback basics, 25
 recording basics, 21
 tape loading, 27, 114
Bilingual programming, 51
Brake, mechanical, 217
Burst ID (Beta), 21

C

Cable ready, 50
Camera sync problems, 81
Capstan:
 circuits, 188
 FG pulses, 13
Carrier phase inverter (Beta), 20
Cassette:
 loading, 45
 load/unload, 127
 up switch, 325
Channel:
 display, 233
 lock (tuner), 234
 scan (AFT), 230
Chroma (color) circuits, 274
Color burst, 22
Color-under system, 15
Comb filter, 17
Communications circuits, 112
Communications microprocessor, 312
Control track pulses, 14
Copy problems, 80
Copyright problems, 64
Counter circuits, 290, 327
Counters, frequency, 66
CTL signal, 11

Current limit circuits, 95
Cylinder, 9
  circuits, 180
  FG pulses, 11
  lock detector, 325
  tack pulses, 13

D

DD (direct-drive), 10
  motors, 49
Delay line, 1H, 17
Delta phi notation, 39
Design alterations, 63
Dew sensor, 144
  circuits, 323
Digital probes, 68
Display circuits, 287, 298
DOC, 25
Dolby, 51, 245
Double-speed circuits, 200
Down-conversion, 15
Dropout, 26
Drum, 9
Dubbing, audio, 5, 57, 244

E

E-E, 24
End sensor, 142
End-of-tape sensors, 82

F

Fast-forward, 141
FG/CTL circuits, 205
fH,fH*, 18
Fine edit, 52
  circuits, 305
Five-head, 165
Fluorescent display circuits, 295
Flutter problems, 81
Forward search, 136, 206
Forward sensor circuits, 323
Four-head, 157, 161
Freeze frame, 165, 195

Frequency counters, 66
Frequency synthesized tuning, 50
Front-load:
  circuits, 149
  mechanics, 123
FS tuning circuits, 239

G

Generators, signal, 64
Guard band, 15

H

Hall-effect circuits, 299
Handling (VCRs), 62
Heads:
  cleaning, 72
  multiple, 49
  selection, 154
  stereo, 244
  switching, 155
Helical scan, 9

I

Independent two-channel
    programming, 51
Index circuits, 305
Interchange problems, 83
Interlace problems, 81
Interlocks, 63
IR remote control, 100

J

Jitter problems, 217

K

Keyboard circuits, 287
Key matrix, 228

L

Leakage current tests, 61
Line period (1H), 11
Loading:
  Beta tape, 31, 114
  motor, 144
  VHS tape, 45, 131
Luminance (B/W) circuits, 273

M

Malerase function, 49
Mechanical brake, 217
Meters, 66
Microprocessor, system control, 314
Moisture sensor, 144
  circuits, 323
Monitor control, 51
Monitors, TV, 68

N

Noise, background, 245
Noise bars, 188

O

ON-OFF circuits, 318
Oscilloscopes, 66
OTR, 52, 283

P

Phase-inversion recording, 16, 30
Playback basics:
  Beta, 25
  VHS, 34
PLL, 7
PLL microprocessor communication, 238
PLL/tuner circuits, 242
Position modulation, 100
Power failure protection, 139
Probes, 68
Programming circuits, 280
Pulse-swallow circuits, 224

Q

Quasi-sync circuits, 219

R

Recording basics:
  Beta, 21
  VHS, 32
Record lockout function, 49
Record time circuits, 327
Rectifier circuits, 91
Reel motor circuits, 134, 147
Reel sensors, 324
REF 30 pulses, 13
Remaining tape circuits, 292
Remote control, 50
  receiver circuits, 108
  transmitter circuits, 103
Reverse search, 137, 207
Rewind brake, 137
Rewind sensor circuits, 323
RF unit problems, 80

S

Safety, 59
Safety notices, 63
Scanner, 9
Sensor circuits, 142, 323
Servo, 10, 83
Signal generators, 65
Slack tape sensors, 82, 144
Slow-motion, 194
  continuously variable, 215
Slow-still:
  circuits, 192
  playback, 166
Slow-tracking circuits, 112
Start sensor, 142

Stereo, 51, 244
Still circuit, 195, 198
Still field head, 165
Stop sensor, 142
Storage, VCR, 62
Supply drive, 47
Supply motor, 139
Switched-B+ supplies, 88
System control circuits, 322
System control problems, 82
Sync, camera, 81
Sync, tuner, 233

Tracking error, 155
Trouble sensor circuits, 325
Troubleshooting approach, 74
Tuner:
  B+ circuits, 229
  PLL circuits, 242
  prescaler, 227
  problems, 80
TV AFC compatability problems, 79
TV/VCR select circuits, 110

V

Vertical jitter problems, 217
VHS:
  playback basics, 34
  recording basics, 32
  tape loading, 45, 131
Video fine edit, 52
Video monitors, 68
Voltage regulator circuits, 93
V-V, 24

T

Take-up drive, 47
Take-up motor, 135
Tape:
  alignment, 69
  calculator, 299
  cleaning, 73
  counter, 290, 301
  loading, 31, 45, 114, 131
  transport sensors, 142
Tension sensor, 139
Timer circuits, 279, 287
Time-remaining circuits, 302
Tools, 71
Top-load mechanics, 115

W

Warning symbols, 59
Wow problems, 81